EDP SCIENCES PROCEEDINGS

Statistics for Astrophysics
Bayesian Methodology

Didier Fraix-Burnet, Stéphane Girard,
Julyan Arbel and Jean-Baptiste Marquette, Eds

edp sciences

ISBN(print): 978-2-7598-2274-4 – ISBN(ebook): 978-2-7598-2275-1

Organisers

Fraix-Burnet Didier, Univ. Grenoble Alpes, CNRS, IPAG, France
Girard Stéphane, Univ. Grenoble Alpes, Inria, CNRS, LJK, Grenoble, France
Arbel Julyan, Univ. Grenoble Alpes, Inria, CNRS, LJK, Grenoble, France
Marquette Jean-Baptiste, LAB, CNRS, Univ. de Bordeaux, France

Lecturers

van Dyk David, Imperial College London, UK
Robert Christian, Univ. Paris-Dauphine, Univ. Warwick, France/UK
Stennig David, Imperial College London, UK
Xixi Yu, Imperial College London, UK
Arbel Julyan, Univ. Grenoble Alpes, Inria, CNRS, LJK, Grenoble, France

Acknowledgments

We thank our sponsors:

- Labex OSUG@2020, Grenoble

- Maison de la Modélisation et de la Simulation, Nanosciences et Environnement (Maimosine, Grenoble)

- Institut de Planétologie et d'Astrophysique de Grenoble (IPAG)

- Idex - Comue Université Grenoble Alpes

- Inria Grenoble Rhône-Alpes

- Laboratoire Jean Kuntzmann (LJK)

- Université Grenoble Alpes

- Grenoble INP

- Programme National de Cosmologie et Galaxies (PNCG), Institut National des Sciences de l'Univers (INSU), CNRS

and the support from the

- GdR MaDICS (CNRS)

- International Society for Bayesian Analysis (ISBA).

List of Participants

Berihuete Angel, Universidad de Cadiz, Spain
Bernard Edouard, Observatoire de la Côte d'Azur, France
Bernus Léo, IMCCE, France
Bilton Lawrence, The University of Hull, UK
Bontemps Sylvain, Laboratoire d'Astrophysique de Bordeaux, France
Boux Fabien, University Grenoble Alpes, France
Buss Claudia, Laboratoire d'Astrophysique de Marseille, France
Cai Yongzhi, University of Padova, Italy
Cornu David, UTINAM, France
Dauvergne Frédéric, Paris-Meudon Observatory, France
Forbes Florence, Inria, France
Galli Phillip, University of Sao Paulo, Brazil
Hackstein Stefan, Sternwarte Bergedorf, Germany
Harrison Ian, University of Manchester, UK
Hestroffer Daniel, IMCCE/Paris Observatory, France
Hottier Clément, Observatoire de Paris, France
Khalaj Pouria, Inst. of Planetology and Astrophysics of Grenoble, France
Khusanova Yana, Laboratoire d'Astrophysique de Marseille, France
Kovlakas Konstantinos, University of Crete IESL/FORTH, Greece
Kruuse Maarja, Tartu Observatory, Estony
Lalande Florian, ENSAI / ERC COSMIC DANCE, France
Law Chi Yan, Chinese university of Hong kong, Hong Kong
Lescinskaité Alina, Center for Physical Sciences and Technology, Lituany
Lestrade Jean-Francois, LERMA Observatoire de Paris, France
Lu Hongliang, Inria, France
Marshall Douglas, Université Paris Diderot, France
Miret Roig Nuria, Université de Bordeaux, France
Munoz Ramirez Veronica, INSERM, France
Osborn Hugh, Laboratoire d'Astrophysique de Marseille, France
Pagani Laurent, CNRS Observatoire de Paris, France
Pascale Raffaele, University of Bologna, Italy
Pilia Maura, INAF - Osservatorio di Cagliari, Italy
Piron Frédéric, CNRS/IN2P3/LUPM, France
Ramos Ramirez Pau, Barcelona University, Spain
Sagredo Bryan, Universidad de Chile, Chile
Setyawati Yoshinta, AEI Hannover, Germany
Slezak Éric, Observatoire de la Côte d'Azur, France
Spoto Federica, Observatoire de la Côte d'Azur, France
Terrazas Vargas Juan Carlos, Universidad de Chile, Chile
Thiel Vivien, Max Planck Institut fuer Radioastronomie, Germany
Vega Garcia Laura, Max Planck Institut fuer Radioastronomie, Germany
Vono Maxime, IRIT, France

Table of Contents

Foreword   9

BAYESIAN STATISTICAL METHODS FOR ASTRONOMY
PART I: FOUNDATIONS
        David C. Stenning and David A. van Dyk   11

1 Foundations of Bayesian Data Analysis   11
2 Further Topics with Univariate Parameter Models   22
3 Final Comments   26
References   27

BAYESIAN STATISTICAL METHODS FOR ASTRONOMY
PART II: MARKOV CHAIN MONTE CARLO
        David C. Stenning and David A. van Dyk   29

1 Introduction   29
2 Rejection Sampling   30
3 Markov Chain Monte Carlo   34
4 Practical Challenges and Advice   42
5 Overview of Recommended Strategy   55
References   56

BAYESIAN STATISTICAL METHODS FOR ASTRONOMY
PART III: MODEL BUILDING
        David C. Stenning and David A. van Dyk   59

1 Introduction to Multi-Level Models   59
2 A Multilevel Model for Selection Effects   60
3 James-Stein Estimators and Shrinkage   66
4 Hierarchical Models and the Bayesian Perspective   70
5 Concluding Remarks   73
References   75

APPROXIMATE BAYESIAN COMPUTATION, AN INTRODUCTION
            Christian P. Robert   **77**

1 Mudmap: ABC at a glance   77
2 ABC Basics   80
3 ABC Consistency   93
4 Summary Statistics, the ABC Conundrum   96
5 ABC Model Choice   98
6 Conclusion   107
References   107

**CLUSTERING MILKY WAY'S GLOBULAR CLUSTERS: A BAYESIAN NONPARAMETRIC
APPROACH**
            Julyan Arbel   **113**

1 R requirements   113
2 Introduction and motivation   114
3 Model-based clustering   117
4 Bayesian nonparametrics around the Dirichlet process   119
5 Application to clustering of globulars of our galaxy   130
References   137

Foreword

This book is the result of the third session of the School of Statistics for Astrophysics that took place on October, 9 to 13, 2017, at Autrans near Grenoble, in France. It was devoted to the Bayesian Methodology and gathered 43 participants. Among statistical approaches used in astrophysics, Bayesian methods become quite famous and popular. They are particularly developed in cosmology. But are they always understood and implemented correctly? Since the first session of the School of Statistics for Astrophysics in 2013, many tutorial sessions and schools intended to train astronomers to modern statistics have spread out around the world. However, they are rarely focused on in depth courses on different aspects of the Bayesian methodology for a full week. Our lecturers are highly recognised statistician experts in this field, while regularly collaborating with astronomers. The participants found the level of the courses somewhat difficult, but it is probably related to the intrinsic complexity of the Bayesian approach itself.

The first chapter provides a detailed introduction to the Bayesian philosophy and its justification. In particular, it makes a comparison with other methods using likelihood. It insists on the particularity of the Bayesian approach, which is the introduction of a prior distribution. This prior is necessarily subjective and may have some influence on the result, but it can bring interesting improvements over likelihood-based methods.

The surge of the Bayesian methodology in science in general has been made possible by the somewhat recent computational power. One of the most heavily used techniques, at least in astronomy, is the Markov Chain Monte Carlo algorithm. The second chapter is thus devoted to the modern statistical computing techniques that are compulsory in any Bayesian analysis.

The third chapter presents the fundamental step of the definition of prior distributions for the unknown (statistical or physical) model parameters. The so-called model building is often multi-level, leading to the classes of hierarchical Bayesian analyses commonly used in cosmology.

When the likelihood cannot be evaluated, the Approximate Bayesian Computation comes to the rescue. This can be seen as a nonparametric inference approach, and is the subject of very active current research. The fourth chapter introduces the basis of the ABC techniques and gives an overview of recent developments.

Finally, the fifth chapter presents a Bayesian nonparametric approach to clustering (unsupervised classification, see the second session of our School of Statistics for Astrophysics[1] held in 2015). Many clustering methods, especially the partitioning ones, require the number of clusters to be initially guessed. This is however not always possible, and the Bayesian nonpara-

---

[1]http://stat4astro2015.sciencesconf.org/

metric approach can thus be interesting in this context.

Many books on Bayesian methodology exist, but they are either too detailed for an astronomer, or too basic and thus not enough for a serious application. We hope this book, that also contains practical exercises, will provide a useful intermediate. In addition, this book should make it clear that statistics is a very active field, and that Bayesian methodology cannot be considered as a traditional, robust and definitive black box. Each case is unique, and hidden caveats are always present.

The reader can find on the school website `http://stat4astro2017.sciencesconf.org/` some complementary information (such as an introduction to R as well as some codes and data sets).

Didier Fraix-Burnet
Stéphane Girard
Julyan Arbel
Jean-Baptiste Marquette

BAYESIAN STATISTICAL METHODS FOR ASTRONOMY
PART I: FOUNDATIONS

David C. Stenning and David A. van Dyk [1]

Abstract. The use of statistical methods in general and Bayesian methods in particular is growing exponentially in astronomy. Bayesian methods allow astronomers to directly model the complexities of sources and instruments, and perform science-driven modeling as opposed to just predictive modeling. The clear mathematical foundations on which Bayesian methods are based allow hierarchical or multi-level structure in data/models, and provide a straightforward way to combine multiple information sources and/or data streams. The main challenge is that Bayesian methods require us to specify "prior distributions" on unknown model parameters. This chapter will introduce the foundations of data analysis from a Bayesian perspective, using examples from astronomy to clarify the mathematical concepts.

1 Foundations of Bayesian Data Analysis

1.1 Probability

What do we mean when we consider the probability of rolling two dice and getting doubles? Or the probability that it will rain today? How should we define "probability?" It is not uncommon to rely on repeatable events such as identical experiments and define the probability of an outcome based on how frequently it occurs; this is the *frequentist definition* of probability. For example, when we flip a coin we say the probability of it coming up heads is 50% if over many trials the proportion of heads is one-half. Astronomy and astrophysics, however, typically do not allow for identical/repeated experiments; there is only one universe we can observe! An alternative definition of probability is that it is a measure of personal certainty, information, or knowledge about the likelihood of the occurrence of an event. That is, probability is subjective and you and I may have different probabilities for the same event; this is the *subjective definition* of probability. For example, prior to flipping a coin we may say that there is a 50% probability of it coming up heads because we have expectations based on our knowledge of coin flipping,

[1] Statistics Section, Department of Mathematics, Imperial College London

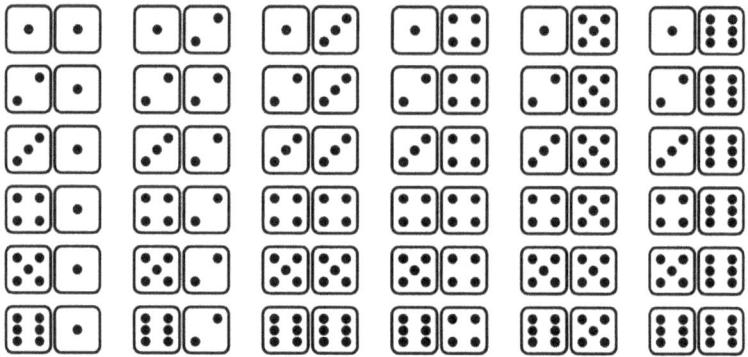

Fig. 1. The set of possible outcomes of rolling two dice.

or because there are two possible outcomes and we weight them equally in the absence of other information. If we truly adhere to the frequentist definition of probability, we cannot make probabilistic claims about coin flips unless we perform the experiment many times. Further, probabilistic statements about non-repeatable events, such as the possibility of rain on a particular day, necessitate the subjective interpretation.

While the frequentist and subjective definitions of probability may seem inconsistent or contradictory, they merely aim to interpret a *probability function*. Mathematically, a probability function is a function that we use to quantify the probability of possible events. Before delving into Bayesian methods, it is helpful to be more formal about exactly what we mean by a probability function.

Suppose we roll two dice and let S be the set of possible outcomes, which is displayed in Figure 1. We call S the *sample space*, which is the set of all possible outcomes of a statistical experiment. An *event A* is any subset of the sample space, $A \subseteq S$; that is, A is a collection of *some* of the possible outcomes of the experiment. For example, $A = \{$two sixes$\} = \{$⚅⚅$\}$, or $A = \{$both faces the same$\} = \{$⚀⚀, ⚁⚁, ⚂⚂, ⚃⚃, ⚄⚄, ⚅⚅$\}$. A probability function is mathematically defined so that the probability of the sample space is one and the probability of any event is non-negative and the sum of the probabilities of its disjoint subsets. In this way a probability function quantifies probabilities, but does not specify how to interpret them in a "real-world" problem.

The formal mathematical definition of probability relies on the *Kolmogorov Axioms*, which state that a probability function is a function such that

i) $\Pr(A) \geq 0$, for all subsets of S.

ii) $\Pr(S) = 1$.

iii) For any pair of disjoint subsets, A_1 and A_2, of S, $\Pr(A_1 \text{ or } A_2) = \Pr(A_1) + \Pr(A_2)$.[1]

---

[1](Countable additivity) More precisely, if $A_1, A_2, \ldots$ are pairwise disjoint subsets of S then

All of the standard properties of probability functions follow from these axioms. The rest of this chapter assumes familiarity with the following:

1. Probability mass functions for discrete random variables, e.g.,

    i. $\Pr(a \leq X \leq b) = \sum_{x=a}^{b} p_X(x)$

    ii. $\sum_{x \in \mathbb{X}} p_X(x) = 1$, where $\mathbb{X}$ is the range of the random variable.

2. Probability density functions for continuous random variables, e.g.,

    i. $\Pr(a < X < b) = \int_a^b p_X(x)dx$

    ii. $\int_{-\infty}^{\infty} p_X(x)dx = 1$.

3. Joint probability functions, e.g.,

    i. $\Pr(a < X < b \text{ and } Y > c) = \int_a^b \int_c^\infty p_{XY}(x,y)dydx$

    ii. $p_X(x) = \int_{-\infty}^\infty p_{XY}(x,y)dy$.

4. Conditional probability functions, e.g.,

    i. $p_Y(y|x) = p_{XY}(x,y)/p_X(x)$

    ii. $p_{XY}(x,y) = p_X(x)p_Y(y|x)$.

Note that when it is clear from context, we omit the subscripts: $p(x) = p_X(x)$. The concepts of expectation, $E(X)$, and variance, $\mathrm{Var}(X)$, should also be familiar. Expectation and variance measure the center and spread of a distribution, respectively. For a continuous random variable X, $E(X)$ and $\mathrm{Var}(X)$ are calculated using the following formulas:

$$E(X) = \int_{-\infty}^{\infty} x p_X(x)dx$$

$$\mathrm{Var}(X) = E(X^2) - [E(X)]^2,$$

where $E(X^2) = \int_{-\infty}^\infty x^2 p_X(x)dx$.

A particularly useful theorem at the heart of Bayesian methods is *Bayes' Theorem*:

$$p_Y(y|x) = \frac{p_X(x|y)p_Y(y)}{p_X(x)} \propto p_X(x|y)p_Y(y).$$

Bayes' Theorem is a powerful tool because it allows us to reverse a conditional probability. It follows from applying the definition of conditional probability twice:

$$p_Y(y|x) = \frac{p_{XY}(x,y)}{p_X(x)} = \frac{p_X(x|y)p_Y(y)}{p_X(x)} \propto p_X(x|y)p_Y(y).$$

The denominator, sometimes called the "evidence" in the astronomy literature, does not depend on y and thus can be viewed as a normalizing constant for the (conditional) probability density function of y (given x). This is advantageous for many computational techniques, as we shall see in later chapters.

$\Pr\left(\bigcup_{i=1}^{\infty} A_i\right) = \sum_{i=1}^{\infty} \Pr(A_i)$.

1.2 Bayesian Analysis of a Standard Poisson Model

As an example, we consider a Poisson model for a photon counting detector. The simplest case is a single-bin detector:

$$Y \overset{\text{dist}}{\sim} \text{POISSON}(\lambda_S \tau),$$

where Y is a discrete random variable for the photon count, τ is the observation time in seconds, and λ_S is the expected counts per second. The *sampling distribution* is the probability function of the data, in this case the observed photon count y:

$$p_Y(y|\lambda_S) = \frac{e^{-\lambda_S \tau}(\lambda_S \tau)^y}{y!}.$$

The *likelihood function* is the sampling distribution viewed as a function of the parameter, where constant factors may be omitted. The likelihood function is the primary statistical tool for assessing the viability of a parameter value given the observed data and a proposed statistical model. Parameters are typically estimated by the *maximum likelihood estimate* (MLE)—the value of the parameter that maximizes the likelihood. For our single-bin detector,

$$\text{likelihood}(\lambda_S) = \frac{e^{-\lambda_S \tau}(\lambda_S \tau)^y}{y!} \qquad \text{and} \qquad \text{log-likelihood}(\lambda_S) = -\lambda_S \tau + y \log(\lambda_S). \quad (1.1)$$

Suppose $y = 3$ with $\tau = 1$. Taking the derivative of the log-likelihood with respect to λ_S and setting it equal to zero, it is easy to show that the MLE is $\hat{\lambda}_S = y/\tau = 3$. (Evaluating the second derivative at the MLE verifies that it is indeed a maximum.) Thus an estimate of λ_S is $\hat{\lambda}_S = 3$. Error bars, which quantify the uncertainty of the estimate, can be calculated using a *normal approximation*—the probability density function of a normal (i.e. Gaussian) distribution with the same mode and curvature as the likelihood function. Specifically, because the mean and variance of a $\text{POISSON}(\lambda_S \tau)$ random variable are both $\lambda_S \tau$, the normal approximation is $\text{NORM}(\mu = \hat{\lambda}_S \tau, \sigma^2 = \hat{\lambda}_S \tau)$, where $\text{NORM}(\mu, \sigma^2)$ is the normal probability density function with mean μ and variance σ^2. The quality of these estimates depend on how well the likelihood function matches its normal approximation; these are illustrated in Figure 2.

A key advantage to Bayesian methods is the ability to incorporate data-appropriate models and methods. Many common statistical methods used in astronomy are based on χ^2 or Gaussian assumptions that may not be well suited to a particular analysis. Bayesian/likelihood methods, on the other hand, easily incorporate more appropriate distributions. For example, for count data we use a Poisson likelihood, which has log-likelihood

$$-\sum_{\text{bins}} \lambda_i + \sum_{\text{bins}} y_i \log \lambda_i,$$

where y_i is the counts in bin i and λ_i its expectation. We can contrast this with typical χ^2 fitting:

$$-\sum_{\text{bins}} \frac{(y_i - \lambda_i)^2}{\sigma_i^2}$$

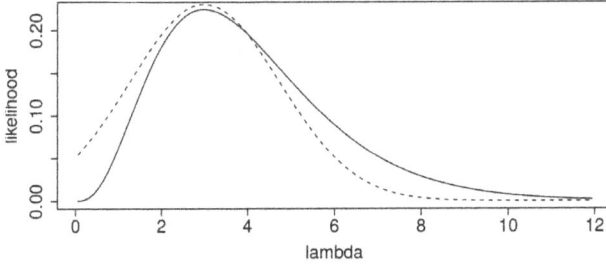

Fig. 2. The likelihood (solid curve) and its normal approximation (dashed curve) for the single-bin detector with $y = 3$ and $\tau = 1$. The normal approximation is the normal probability density function with the same mean and curvature as the likelihood function.

or a Gaussian log-likelihood:

$$-\sum_{\text{bins}} \sigma_i - \sum_{\text{bins}} \frac{(y_i - \lambda_i)^2}{\sigma_i^2},$$

where σ_i^2 is the variance, perhaps known through measurement error, in bin i.

After constructing the likelihood function, to perform a Bayesian analysis we must specify a *prior distribution* on the unknown model parameters, which quantifies knowledge regarding the parameters obtained prior to the current observation. For our Poisson model a reasonable choice for the prior distribution on λ_S is the *gamma distribution*, which has probability density function

$$p(\lambda_S) = \frac{\beta^\alpha}{\Gamma(\alpha)} \lambda_S^{\alpha-1} e^{-\beta\lambda_S} \tag{1.2}$$

for $\lambda_S > 0$ and $\alpha, \beta > 0$. Here, α and β are *hyperparameters*—parameters of the prior distribution that may or may not be of scientific interest. With this choice of prior distribution, $E(\lambda_S) = \alpha/\beta$ and $\text{Var}(\lambda_S) = \alpha/\beta^2$. Different choices for α and β yield different densities, which makes the gamma distribution a flexible family of prior distributions for intensity parameters; see Figure 3.

The *posterior distribution* quantifies combined knowledge for parameters obtained *prior* to and with the current observation. We can derive the posterior distribution for the single-bin detector Poisson model via Bayes' Theorem. The posterior distribution is

$$
\begin{aligned}
p(\lambda_S|y) &= p(y|\lambda_S)p(\lambda_S)/p(y) \\
\text{posterior}(\lambda_S|y) &\propto \text{likelihood}(\lambda_S|y) \times p(\lambda_S) \\
&\propto \frac{(\lambda_S\tau)^y e^{-\lambda_S\tau}}{y!} \times \frac{\beta^\alpha}{\Gamma(\alpha)} \lambda_S^{\alpha-1} e^{-\beta\lambda_S} \\
&\propto \lambda_S^y e^{-\lambda_S\tau} \times \lambda_S^{\alpha-1} e^{-\beta\lambda_S} \\
&\propto \lambda_S^{y+\alpha+1} e^{-(\tau+\beta)\lambda_S} .
\end{aligned}
\tag{1.3}
$$

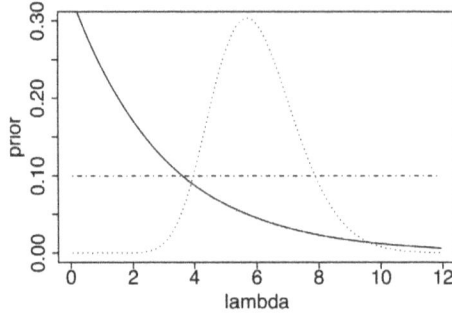

Fig. 3. The gamma distribution provides a flexible family of prior distributions for intensity parameters. The solid, dotted, and dot-dashed lines are the gamma probability density function for three different combinations of α and β; see Equation (1.2).

Comparing (1.2) and (1.3), we identify (1.3) as an (unnormalized) gamma probability density with parameters $y + \alpha$ and $\tau + \beta$. Thus, including information from the data, our knowledge as to the likely value of λ_S is summarized by $\lambda_S|y \sim \text{GAMMA}(y + \alpha, \beta + \tau)$. The posterior distribution combines past and current information, as seen in Figure 4.

As in this example, if the prior and the posterior distributions are of the same family, the prior distribution is called that likelihood's *conjugate prior distribution*. Specifically, if $Y|\lambda_S \overset{\text{dist}}{\sim} \text{POISSON}(\lambda_S \tau)$ and $\lambda_S \overset{\text{dist}}{\sim} \text{GAMMA}(\alpha, \beta)$ then $\lambda_S|Y \overset{\text{dist}}{\sim} \text{GAMMA}(y + \alpha, \tau + \beta)$, that is, with a Poisson likelihood and gamma prior distribution, the posterior distribution will also be a gamma distribution.

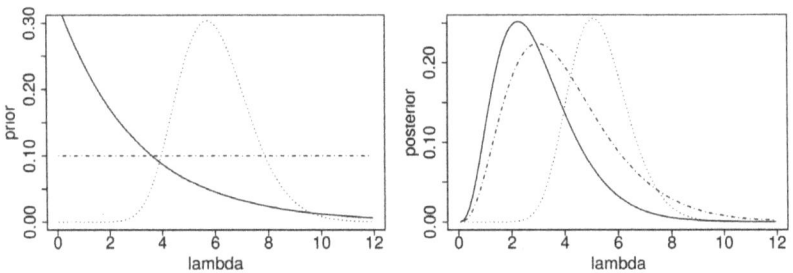

Fig. 4. The posterior distributions (right) for the Poisson model resulting from different choices of prior distribution (left). The three different prior distributions on the left (different line types) correspond to the three posterior distributions on the right (corresponding line type), after having observed y.

While the posterior distribution provides a statistical summary of the unknown model parameter λ_S, researchers would often like simpler summaries such as parameter estimates and error bars. A Bayesian estimator of λ_S is given by the posterior mean

$$E(\lambda_S|y) = \frac{y + \alpha}{\tau + \beta},$$

with Bayesian error bars equal to the posterior standard deviation

$$\overline{\text{Var}(\lambda_S|Y)} = \frac{\sqrt{y + \alpha}}{\tau + \beta}.$$

It is useful to compare the MLE of λ_S,

$$\text{MLE}(\lambda_S) = \frac{y}{\tau},$$

and the posterior expectation of λ_S,

$$E(\lambda_S|y) = \frac{y + \alpha}{\tau + \beta}.$$

The MLE takes into account only the current data, y, and the exposure time, τ. That is, without considering prior or ancillary knowledge, the photon rate λ_S is estimated by the average number of photons recorded per second. The Bayesian estimator, however, incorporates ancillary information vis-à-vis the prior distribution, which has as much influence as α observed counts in an exposure of β seconds. We can use this formulation of the prior distribution in terms of "prior data" to meaningfully specify a prior distribution for λ_S and limit the influence of the prior distribution. Smaller values of β correspond to less prior information.

1.3 Building Blocks of Modern Bayesian Analyses

The first step in a Bayesian analysis is specifying the statistical model. This consists of specification of (i) the likelihood function and (ii) the prior distribution, both of which involve subjective choices. A comprehensive description of the scientific mechanisms can be overly complex and we often rely on parsimony: simple descriptions that do not compromise scientific objectives. It is often helpful to consider what we model versus what we consider to be fixed. For example, a researcher might apply some preprocessing to the data that depends on a calibration product such as an effective area curve. Should calibration uncertainty be considered a component of the statistical model? A common aphorism attributed to the statistician George Box is that "all models are wrong, but some are useful" (Box, 1976). We keep this principle in mind when constructing more complex statistical models.

For an extension of our single-bin detector example, consider that there is background contamination. Let the contaminated source counts be $y = y_S + y_B$, where y_S and y_B are the (unobserved) counts coming from the source and the background, respectively. That is, the observed counts, y, are the sum of the source and background counts and we do not know the proportion. However, assume we can take a background-only observation with an exposure time that is 24 times that of the source exposure, and let the background counts be x. Then a Poisson *Multi-Level Model* for this example is given by

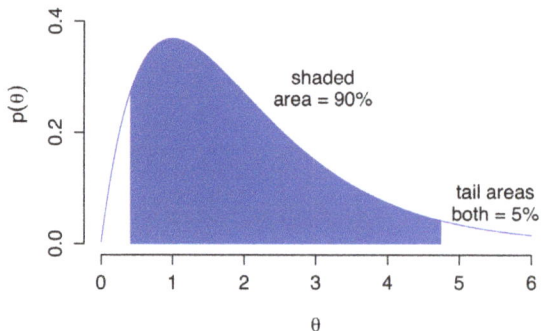

Fig. 5. An equal-tailed interval for a gamma distribution. Equal-tailed intervals have equal probability (i.e. area) in the left and right tails of the distribution and always include the median.

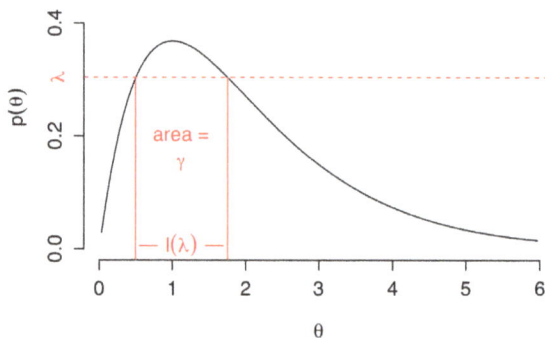

Fig. 6. An HPD interval for a gamma distribution. As λ decreases, the area below the probability density curve and between the two vertical lines, γ, increases. But γ is the probability associated with the interval $I(\lambda)$. HPD intervals contain values of the parameter that have the highest probability density and always include the mode.

Level 1: $y|y_B, \lambda_S \overset{\text{dist}}{\sim} \text{POISSON}(\lambda_S) + y_B$

Level 2: $y_B|\lambda_B \overset{\text{dist}}{\sim} \text{POISSON}(\lambda_B)$ and $x|\lambda_B \overset{\text{dist}}{\sim} \text{POISSON}(\lambda_B \cdot 24)$

Level 3: specify a prior distribution for λ_B, λ_S.

Notice that each level of the multi-level model specifies a distribution given unobserved quantities whose distributions are given in lower levels. In this way, we are able to use multilevel models to break up a complex model into a sequence of relatively simple models.

1.4 Constructing Error Bars on Unknown Model Parameters

As previously discussed, the full statistical summary of the unknown model parameters is given by the posterior distribution, but researchers often would like simpler summaries such as parameter estimates and error bars. These can be given by posterior means and posterior standard deviations, respectively, but what about non-Gaussian posterior distributions? For such cases, with a generic parameter θ, we can find a lower-limit L and upper-limit U so that

$$\Pr(L < \theta < U|y) = \int_L^U p(\theta|y)d\theta = c,$$

where c is chosen to give a $c \times 100\%$ interval. (Note that a "one sigma" interval would use $c = 0.683$, and a "two sigma" interval would use $c = 0.954$.) Because we have specified a prior distribution on the unknown model parameter(s), we can interpret this *Bayesian credible interval* as there is a $c \times 100\%$ (subjective) probability that the true value of the parameter lies in the interval. (By contrast, a *frequentist confidence interval* means that over a large number of repeated trials, $c \times 100\%$ of such intervals will include the true value of the parameter.) More generally, for a generic parameter vector Θ we calculate

$$\Pr(\theta \in \Theta|y) = \int_{\theta \in \Theta} p(\theta|y)d\theta = c.$$

But the intervals calculated in the way are not unique! Are there optimal choices?

One option is the *Equal-Tailed Interval*, which is the simplest interval to compute (e.g., via Monte Carlo). Equal-tailed intervals have equal probability above and below the interval (i.e. in the tails); see Figure 5 for an example. A nice property of equal-tailed intervals is that they are preserved under monotonic transformations. For example, if (L_θ, U_θ) is a 95% equal-tailed interval for θ, then $\left(\log(L_\theta), \log(U_\theta)\right)$ is a 95% equal-tailed interval for $\log(\theta)$. Another choice of posterior interval is the *Highest Posterior Density (HPD) Interval*, which is constructed to include values of the parameter that have the highest probability density. In Figure 6, for example, as λ decreases, the probability (γ) of interval ($I(\lambda)$) increases. The HPD interval is the shortest interval of a given probability.

The difference between an equal-tailed interval and HPD interval may be pronounced for some distributions. For example, in Figure 7(a) we have an equal-tailed interval for a gamma distribution that exhibits skewness and in Figure 7(b) we have the HPD interval for the same distribution. The difference between the two intervals will be more pronounced for more skewed distributions. An important consideration is that an HPD interval will always include the mode but an equal-tailed interval may not, particularly if the distribution is highly skewed. (An equal-tailed interval will, however, always contain the median.) For a multimodal posterior, HPD may not be an interval but instead a disconnected region; see Figure 8.

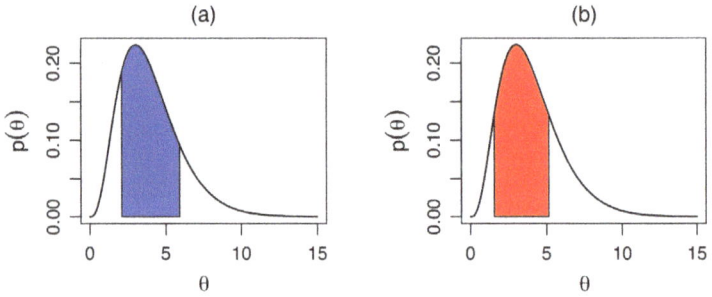

Fig. 7. Comparison of (a) an equal-tailed interval and (b) and HPD interval for a gamma distribution that exhibits skew.

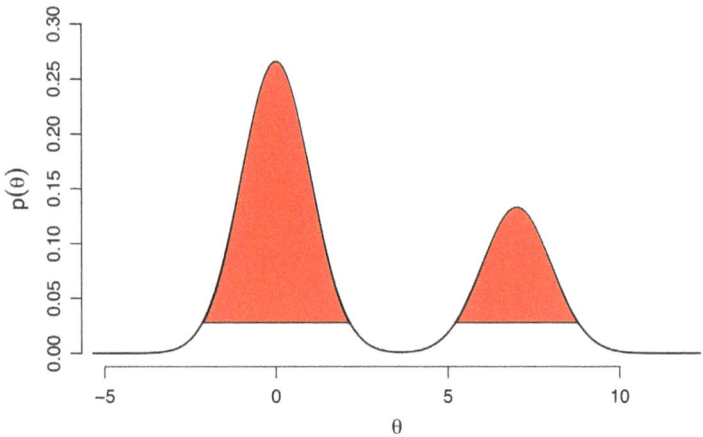

Fig. 8. A disconnected HPD region for a multimodal distribution.

1.5 Comments

Bayesian methods in general benefit from their mathematical foundation. For example, consider the posterior odds for generic parameters θ_1 and θ_2 with generic data y:

$$\frac{p(\theta_1|y)}{p(\theta_2|y)} = \frac{p(y|\theta_1)p(\theta_1)/p(y)}{p(y|\theta_2)p(\theta_2)/p(y)} = \frac{p(y|\theta_1)}{p(y|\theta_2)} \times \frac{p(\theta_1)}{p(\theta_2)}$$

$$= \quad \text{likelihood ratio} \quad \times \quad \text{prior odds}.$$

This can be used to compare two parameter values of interest and is the genesis of Bayesian methods for model comparison. No new methods are required, just standard probability calculations. That is, if the posterior odds in the above formulation is high,

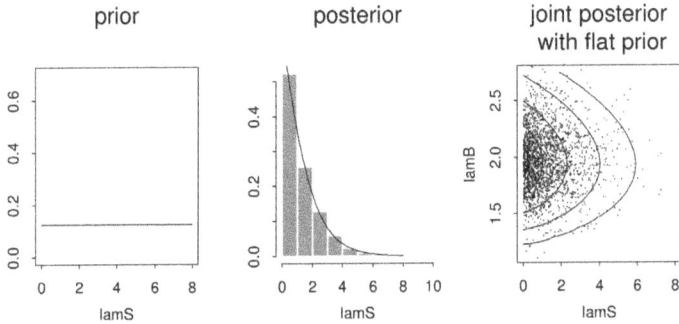

Fig. 9. Flat prior distribution, marginal posterior distribution, and joint posterior distribution for the background- contaminated Poisson model. Solid lines represent mathematically derived distributions, and bars/points a Monte Carlo sample.

then that is evidence in favor of θ_1 over θ_2 after taking into account the current data and our prior knowledge.

For problems with more than one unknown model parameter we have (at least) two ways to summarize the posterior distribution: (i) plot the contours of the posterior distributions of pairs of parameters, and (ii) plot the marginal posterior distributions of the parameters of interest:

$$p(\lambda_S \mid y, y_B) = \int p(\lambda_S, \lambda_B \mid y, y_B) d\lambda_B,$$

where we have integrated out the *nuisance parameter* λ_B. A nuisance parameter is any parameter that is not of scientific interest but which must be included in the statistical model. In this case λ_B, the expected counts per second from the background, is not of immediate scientific concern but must be included in the model because we observe the background-contaminated counts rather than pure source counts.

The posterior distribution, particularly for complex models, is often explored via Monte Carlo methods which easily generalize to higher dimensions. Figure 9 shows the flat prior distribution on λ_S and the resulting marginal posterior distribution, along with the joint posterior distribution of λ_S and λ_B. Solid lines represent the mathematically derived distributions, and bars/points the distributions obtained via Monte Carlo samples. In problems with many unknown parameters, obtaining marginal distributions of individual parameters or of pairs of parameters involves high-dimensional integration. Typically this must be done numerically and quickly becomes intractable as the dimension of the parameter grows. A Monte Carlo sample for the posterior distribution, however, nicely summarizes the posterior while avoiding (potentially costly) numerical integration. Obtaining such Monte Carlo samples is the topic of Chapter 2.

2 Further Topics with Univariate Parameter Models

2.1 Bayesian Analysis of Standard Binomial Model

For another example we consider hardness ratios in high energy astrophysics.[2] Let $H \sim$ Poisson(λ_H) be the observed hard (high energy) count, $S \sim$ Poisson(λ_S) be the observed soft (low energy) count, and $n = H + S$ be the total count. If H and S are independent,

$$H|n \sim \text{Binomial}\left(n, \pi = \frac{\lambda_H}{\lambda_H + \lambda_S}\right).$$

We can conduct a Bayesian analysis of this model, treating π as the unknown parameter. First, we construct the likelihood function:

$$\text{likelihood}(\pi) = p_H(h|\pi) = \frac{n!}{h!(n-h)!}\pi^h(1-\pi)^{n-h} \text{ for } h = 0, 1, \ldots, n.$$

For the prior distribution on π we use a *beta distribution*, which is a flexible class of prior distributions on the unit interval; see Figure 10. The hyperparameters $\alpha, \beta > 1$ define the prior distribution, which has probability density function

$$p(\pi) = \frac{\Gamma(\alpha + \beta)}{\Gamma(\alpha)\Gamma(\beta)}\pi^{\alpha-1}(1-\pi)^{\beta-1} \text{ for } 0 < \pi < 1.$$

Different choices for α and β yield the various distributions in Figure 10.

One reason to use a beta prior distribution for this example is that the beta distribution is conjugate to the binomial. That is, if $H|n, \pi \overset{\text{dist}}{\sim} \text{Binomial}(n, \pi)$ and $\pi \overset{\text{dist}}{\sim} \text{Beta}(\alpha, \beta)$, then

$$\pi|H, n \overset{\text{dist}}{\sim} \text{Beta}(h + \alpha, n - h + \beta).$$

Suppressing the conditioning on n,

$$p(\pi|h) \quad \propto \quad \text{likelihood}(\pi) \times p(\pi)$$

$$= \quad \frac{n!}{h!(n-h)!}\pi^h(1-\pi)^{n-h} \times \frac{\Gamma(\alpha + \beta)}{\Gamma(\alpha)\Gamma(\beta)}\pi^{\alpha-1}(1-\pi)^{\beta-1}$$

$$\propto \quad \pi^{h+\alpha-1}(1-\pi)^{n-h+\beta-1},$$

which as a function of π is proportional to a $\text{Beta}(h+\alpha, n-h+\beta)$ density. A point estimate for π is given by the expectation (i.e., mean) of the beta posterior distribution:

$$\text{E}(\pi|h) = \frac{h + \alpha}{n + \alpha + \beta}.$$

We can interpret the hyper parameters α and β as "prior hard and soft counts." As n increases the choice of prior distribution matters less.

---

[2] For more on Bayesian analysis of hardness ratios see Park et al. (2006).

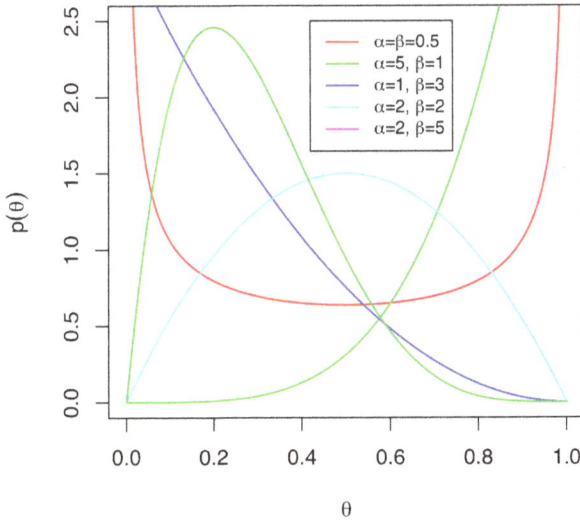

Fig. 10. The beta distribution with different choices for the hyper parameters α and β.

2.2   Transformations

We have formulated our analysis of hardness ratios in terms of

$$\pi = \frac{\lambda_H}{\lambda_H + \lambda_S}.$$

However, other formations are also common:

simple ratio: $\mathcal{R} = \dfrac{\lambda_S}{\lambda_H} = \dfrac{1 - \pi}{\pi}$

color: $C = \log_{10}\left(\dfrac{\lambda_S}{\lambda_H}\right) = \log_{10}(1 - \pi) - \log_{10}(\pi)$

fractional difference: $\mathcal{HR} = \dfrac{\lambda_H - \lambda_S}{\lambda_H + \lambda_S} = 2\pi - 1.$

Transformations of scale and/or parameter are common in statistics, and with a Monte Carlo sample from the posterior distribution transformations are trivial. We simply apply the transformation to every sample to obtain the distribution of the new quantity of interest; see Figure 11.

2.3   Interpreting and Specifying Prior Distributions

In practice, one of the biggest obstacles to implementing a Bayesian analysis is the specification of the prior distribution. In this section we delve deeper into the interpretation

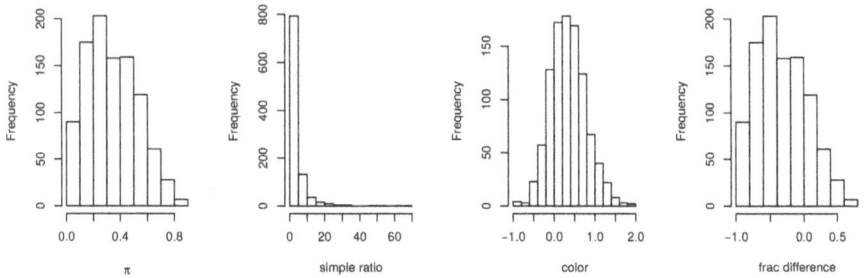

Fig. 11. The posterior distribution of π via a Monte Carlo sample, and the resulting posterior distribution of simple ratio, color, and fraction difference obtained by applying the relevant transformations to the sample.

of a prior distribution with the aim of developing the insight needed to specify a prior in practice.

There are different ways to interpret the prior distribution. Using the hardness ratios for illustration, we have the following three interpretations:

1. POPULATION/FREQUENCY INTERPRETATION: Imagine a population of sources, experiments, or universes from which the current parameter is drawn, i.e., this source is drawn from a population of sources. Here the prior distribution describes the distribution of parameter values across the population.

2. STATE OF KNOWLEDGE: A subjective probability distribution, perhaps based on ancillary scientific information or previous studies.

3. LACK OF KNOWLEDGE: UNIFORM$(0, 1)$ corresponds to "no prior information". This choice of prior distribution does draw $E(\pi|h)$ toward $1/2$, but has relatively large prior variance.

Using the above interpretations we refer to "subjective" and "objective" Bayesian methods. One approach to objective Bayesian analysis relies on *reference priors*—prior distributions that can be used as a matter of course under a given likelihood. That is, once the likelihood is specified the reference prior can be automatically applied. Reference priors might be formulated to (i) minimize the information conveyed by the prior, or to (ii) optimize other statistical properties of estimators. For example, we may find the prior distribution that maximizes $Var(\theta|y)$ or yields confidence intervals with correct frequency coverage.

Another approach to objectivity uses a *non-informative prior*—a prior distribution that aims to play a minimal role in the statistical inference. A common choice for a non-informative prior is a flat or uniform distribution over the range of the parameter. For example, returning to hardness ratios, $h \mid \pi \sim$ BINOMIAL(n, π) with $\pi \sim$ UNIFORM$(0, 1)$ yields a non-informative prior distribution on π. However, a caveat is that while the idea

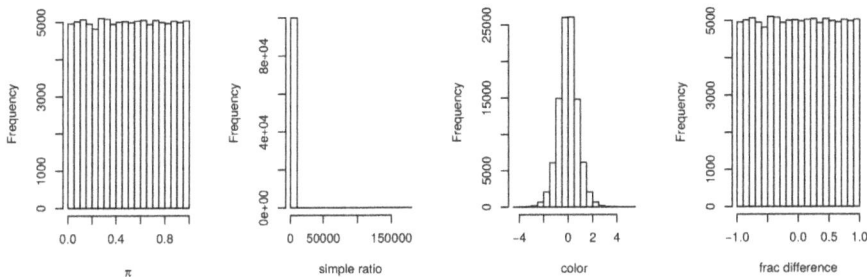

Fig. 12. A flat prior distribution on π, left, does not correspond to a flat prior distribution on simple ratio or color, though it does for fractional difference.

of a "flat prior distribution" seems sensible enough, it is completely determined by the choice of parameterization. Thus a flat prior on π does *not* correspond to a flat prior on simple ratio or color, though it does for fractional difference; see Figure 12. This is because fractional difference is a simple linear transformation of π, whereas the simple ratio and color are not.

One might be tempted to specify a flat prior distribution on ψ = color: $p(\psi) \propto 1$ for $-\infty < \psi < \infty$. This is an example of an *improper prior*—a positive-valued function that is not integrable, but that is used formally as a prior distribution. Improper priors should be used with caution; because they are not distributions we can not rely on probability theory alone. However, improper priors generally cause no problem so long as we verify that the resulting posterior distribution is a proper distribution. If the posterior distribution is *not* proper, no sensible conclusions can be drawn. For example, a flat improper prior distribution on color corresponds to an improper distribution on π: $\pi \sim Beta(\alpha = 0, \beta = 0)$. The posterior distribution, however, is proper so long as there is at least one hard and soft count, i.e., (i) $h \geq 1$ and (ii) $n - h \geq 1$.

As we have seen, a flat non-informative prior on a particular parameterization may be neither flat nor non-informative when using a different parameterization. A sensible question to ask is whether we can find an objective rule for generating priors that do not depend on the choice of parameterization? *Jeffreys' invariance principle* says that any rule for determining a (non-informative) prior distribution should yield the same result if applied to a transformation of the parameter. Any subjective prior distribution should adhere to Jeffreys' invariance principle, at least in principle. In likelihood-based statistics, with generic parameter θ and data y, the *Expected Fisher Information* is

$$J(\theta) = -\mathrm{E}\left[\frac{\mathrm{d}^2 \log p(y|\theta)}{\mathrm{d}^2\theta} \,\middle|\, \theta\right].$$

Using this, the *Jeffreys prior distribution* is

$$p(\theta) \propto \sqrt{J(\theta)},$$

or, in higher dimensions,

$$p(\Theta) \propto \sqrt{|J(\Theta)|},\tag{2.1}$$

where Θ is a generic parameter vector. (In (2.1) $J(\Theta)$ is the opposite of the expected *matrix of partial* derivatives of the log-likelihood function.) For example, for the binomial model

$$\log(p_H(h|\pi)) = h\log(\pi) + (n - h)\log(1 - \pi) + \text{ constant,}$$

the expected Fisher information is

$$-\mathrm{E}\left[-\frac{h}{\pi^2} - \frac{n - h}{(1 - \pi)^2}\,\bigg|\,\pi\right] = \frac{n}{\pi(1 - \pi)},$$

and so the Jeffreys Prior is

$$p(\pi) \propto \sqrt{J(\pi)} \propto \pi^{-1/2}(1 - \pi)^{-1/2} = \text{Beta}(\alpha = 1/2, \beta = 1/2),$$

which is an "arcsine distribution." This prior distribution is non-informative and invariant to reparameterizations.

3   Final Comments

Those concerned about the reliance of Bayesian methods on the choice of prior distribution need not be, at least in most cases. Any reasonable prior distribution results in exactly the same asymptotic frequency properties as likelihood methods. The only worry is if you want to do better than likelihood-based methods in small samples. Nevertheless, in practice much effort is put into selecting priors that help us best achieve our objectives. In fact, an advantage to Bayesian methods is that the choice of prior distribution is an additional degree of freedom in methodological development, and the choice of prior can even improve frequency properties in some cases.

While there may be additional concern related to the supposed subjectivity of Bayesian analyses, it is important to recognize that *all statistical analyses are subjective.* Researchers make choices regarding data (e.g., what data should be collected and any pre-processing that is applied), parametric forms, and statistical/scientific models. Bayesian methods merely have one more subjective component: the quantification of prior knowledge through a distribution on the unknown model parameters. Further, as we have seen, prior distributions need not be used in a subjective manner. The tradeoff for introducing this extra subjective component is that everything follows from basic probability theory once we have established the likelihood function and prior distribution; asymptotic results and counter-intuitive definitions (e.g., for a frequentist confidence interval) are not required.

Readers interested in a more in-depth coverage of Bayesian methods are encouraged to consult Sivia & Skilling (2006) for a reference geared towards the physical sciences, and Gelman et al. (2013) for a more comprehensive overview, among others.

References

Box, G. E.P. (1976). Science and Statistics. *Journal of the American Statistical Association*, 71: 791–799.

Gelman, A., Carlin, J. B., Stern, H. S., Dunson, D. B., Vehtari, A., and Rubin, D. B. *Bayesian Data Analysis: Third Edition* (Chapman & Hall/CRC Press, Boca Raton, London, New York, 2013)

Park, T., Kashyap, V., Siemiginowska, A., van Dyk, D. A., Zezas, A., Heinke, C., and Wargelin, B. J. (2006). Hardness Ratios with Poisson Errors: Modeling and Computations. *The Astrophysical Journal*, 652: 610–628.

Park, T., van Dyk, D. A., and Siemiginowska, A. (2008). Searching for Narrow Emission Lines in X-ray Spectra: Computation and Methods. *The Astrophysical Journal*, 688: 807–825.

Sivia, D.S. and Skilling, J. *Data Analysis: A Bayesian Tutorial: Second Edition* (Oxford University Press, Oxford, 2006)

BAYESIAN STATISTICAL METHODS FOR ASTRONOMY PART II: MARKOV CHAIN MONTE CARLO

David C. Stenning and David A. van Dyk[1]

Abstract. One reason for the surge in Bayesian methods both in general and in astronomy is the sophisticated computational methods that have become available. As computer speed has increased, so has the development of methodologies such as Markov chain Monte Carlo that leverage this speed. Such methodologies allow researchers to fit complex models that would have been beyond consideration not long ago. This chapter provides an introduction to modern statistical computing, focusing on Markov chain Monte Carlo methods in particular. The approaches are explained using examples from astronomy.

1 Introduction

We can use a Monte Carlo (MC) sample from a distribution to derive useful summaries of the distribution. (Here we focus on posterior distributions even though the methods we describe can be applied to any distribution.) If we have an MC sample that is representative of the distribution, we can use the sample mean and variance of each of its component variables to estimate the corresponding marginal means and variances of the distribution; this was demonstrated in Chapter 1 Figure 9. Similarly, we can compute correlations among the variables or make scatter plots to explore complex non-linear relationships among the variables. Further, the replicate values of the parameter in the MC sample need not be independent. Common methods involve using Markov chains to produce a *dependent* Monte Carlo sample from the distribution, a technique known as Markov chain Monte Carlo (MCMC).

For many "real-world" problems it is difficult to directly compute the posterior means, variances, and correlations of the unknown model parameters. For example, Figure 1 illustrates the posterior distribution of the stellar parameters age, metallicity [Fe/H], distance, and zero-age-main-sequence (ZAMS) mass for a particular white dwarf. The dots are an MC sample, and methods for obtaining

---

[1] Statistics Section, Department of Mathematics, Imperial College London

such a sample are the topic of this chapter. The marginal posterior means are given by the blue crosses, where the extent of the crosses in the horizontal and vertical directions represent the respective marginal posterior standard deviations. From this we see that point estimates (i.e., means) and error bars (i.e., standard deviations) are not sufficient for summarizing the posterior due to the highly non-linear posterior relationship among stellar parameters. In fact, examining the posterior distribution of age and distance in the top left panel of Figure 1, we notice that simply relying on marginal posterior means results in estimates that are not (jointly) favored by the model due to the non-linear correlations; that is, there are very few MC samples at the intersection point of the blue cross, which means that there is low posterior density in that region of parameter space. Thus failing to capture the complex correlations among model parameters can result in sacrificing valuable scientific information about the problem at hand and, more-over, yield misleading or incorrect inferences. This is another advantage of MC methods. The acquired MC sample can provide a concise summary of a complex posterior distribution and describe non-linear relationships among variables.

The rest of this chapter proceeds as follows. In Section 2 we discuss a method that produces an independent MC sample: *rejection sampling*. Markov chains are introduced in Section 2.1, and algorithms that use Markov chains to obtain an MC sample, i.e., MCMC algorithms, are described in Section 3. Practical advice on the methodologies introduced in this chapter is provided in Section 4, and a recommended strategy for obtaining MC samples from a distribution is given in Section 5.

2   Rejection Sampling

Consider the distribution $f(\theta)$, for a generic univariate parameter θ, illustrated via the dashed line in Figure 2. Suppose that we wish to acquire a Monte Carlo sample from $f(\theta)$, i.e., $f(\theta)$ is the *target distribution*, but cannot do so directly. Rejection sampling allows us to use a different distribution $g(\theta)$ such that

$$f(\theta) \leq Mg(\theta) \tag{2.1}$$

for some known constant M, with $1 < M < \infty$, to obtain a sample from $f(\theta)$. The requirement (2.1) stipulates that we can *bound* $f(\theta)$ with some unnormalized density, $Mg(\theta)$. In Figure 2, $Mg(\theta)$ is a uniform distribution as illustrated by the red box.

Rejection sampling procedes for $t = 1, 2, \ldots$

Step 1. Sample $\tilde{\theta}$ from the distribution $g(\theta)$: $\tilde{\theta} \sim g(\theta)$.

Step 2. Sample a value u from the unit interval: $u \sim \text{UNIFORM}(0, 1)$.

Step 3. If

$$u \leq \frac{f(\tilde{\theta})}{Mg(\tilde{\theta})}, \text{ i.e., if } uMg(\tilde{\theta}) \leq f(\tilde{\theta}),$$

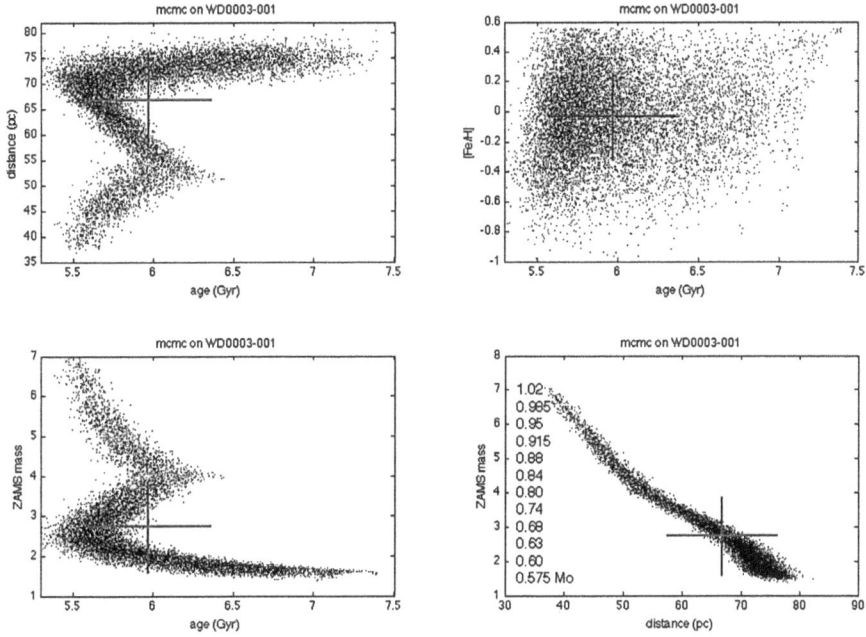

Fig. 1. A complex posterior distribution. The parameters are physical properties of a white dwarf star: age, distance, metallicity [Fe/H], and zero-age-main-sequence (ZAMS) mass. The dots represent a Monte Carlo sample from the posterior, and the blue crosses represent marginal posterior means and standard deviations. Notice that when there are highly non-linear relationships among stellar parameters, the point estimates (i.e., posterior means) and error bars (i.e., posterior standard deviations) fail to adequately summarize the posterior distribution; important scientific information may be lost if we do not learn about the contours of the posterior distribution.

accept $\tilde{\theta}$. That is, set $\theta^{(t)} = \tilde{\theta}$ and proceed to the next iteration. Otherwise reject $\tilde{\theta}$ and return to step 1.

This is illustrated in Figure 2. In this example, treating the values drawn in steps 1 and 2 as the horizontal and vertical coordinates, respectively, is equivalent to drawing uniformly within the red box. We accept a draw if it falls below the dashed density function. After repeating for a total of N iterations, the result is a sample of N independent draws from $f(\theta)$: $\{\theta^{(1)}, \theta^{(2)}, \ldots, \theta^{(N)}\}$,where $\theta^{(t)}$ is the value for θ drawn on iteration t. The sample $\{\theta^{(1)}, \theta^{(2)}, \ldots, \theta^{(N)}\}$ characterizes the distribution $f(\theta)$, and estimates, error bars and uncertainty intervals for the unknown θ can be calculated from the mean, standard deviation and quantiles of

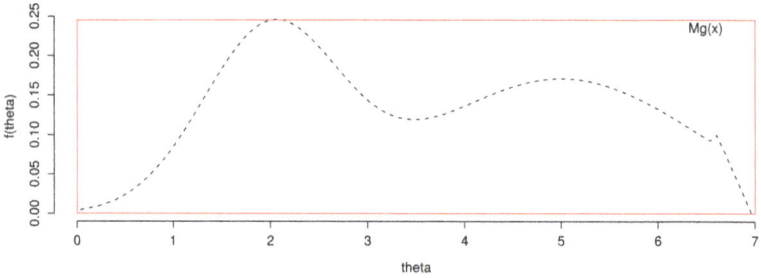

Fig. 2. Illustration for a rejection sampler. We can obtain a sample from the dashed target distribution $f(\theta)$ by drawing values of θ uniformly in the red rectangle and keeping only those that fall below the dashed curve. The result is an independent sample $\{\theta^{(1)}, \theta^{(2)}, \ldots, \theta^{(N)}\}$ from $f(\theta)$, which we can use to make inferences about θ.

the sample, respectively. That is, an estimate of θ is

$$\bar{\theta} = \frac{1}{N} \sum_{t=1}^{N} \theta^{(t)},$$

with standard-deviation based error bar calculated by

$$\theta_s = \frac{1}{N-1} \sqrt{\sum_{t=1}^{N} \left(\theta^{(t)} - \bar{\theta}\right)^2},$$

and a 95% equal-tailed interval for θ is estimated by the interval extending from the 2.5 and 97.5 percentiles of $\{\theta^{(1)}, \ldots, \theta^{(N)}\}$. An error bar based on the standard deviation should only be used for posterior distributions that are roughly symmetric and bell-shaped. The shape of the distribution can be assessed with a histogram of the MC sample. In practice, histograms and/or scatterplots of the MC sample should always be constructed which may reveal multiple modes, (non-)linear correlations, and other features that caution us against relying on simple numerical summaries such as point estimates and error bars.

Rejection samplers are inefficient if the wait time for acceptance is high. Continuing the example from Figure 2, to reduce the wait for acceptance we can improve $g(\theta)$ as an approximation to $f(\theta)$. For example, the blue curve in Figure 3 is a better approximation to $f(\theta)$ than the uniform distribution represented by the red line. The closer $g(\theta)$ is to $f(\theta)$ the more efficient rejection sampling becomes, which is intuitive because if $g(\theta)$ were equal to $f(\theta)$ every draw would be accepted. (But of course we would not need to perform rejection sampling if it is possible to directly sample from $f(\theta)$!)

A major drawback of rejection sampling is that it can be very inefficient in high dimensions due to the difficultly of finding a suitable $g(\theta)$ that does not lead

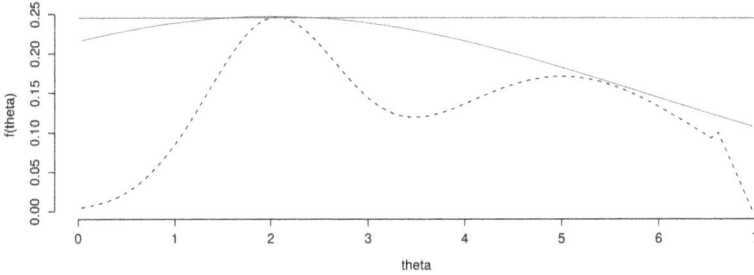

Fig. 3. Two choices for $g(\theta)$ in a rejection sampling algorithm, represented by the red and blue curves. Using the blue curve is more efficient because it is a better approximation to the dashed target distribution.

to many rejections. Instead, MCMC methods may be used. We briefly discuss the theory behind Markov chains in Section 2.1 before introducing MCMC in Section 3.

2.1 Markov Chains

A *Markov chain* is a sequence of random variables,

$$\theta^{(0)}, \theta^{(1)}, \theta^{(2)}, \ldots$$

such that

$$p(\theta^{(t)}|\theta^{(t-1)}, \theta^{(t-2)}, \ldots, \theta^{(0)}) = p(\theta^{(t)}|\theta^{(t-1)}).$$

That is, the distribution of a value in the chain only depends on the history of the chain through its most recent value. As an example, let $\theta^{(0)} = 0$ and, for $t = 1, 2, \ldots$, let $\theta^{(t)} = \theta^{(t-1)} - 1$ with probability 0.5 and let $\theta^{(t)} = \theta^{(t-1)} + 1$ with probability 0.5. This is an example of a *random walk* on the integers. All of the information in the chain $\{\theta^{(t)}, \ldots, \theta^{(t-1)}\}$ as to the distribution of the next value, $\theta^{(t)}$, is contained in its immediate predecessor, $\theta^{(t-1)}$.

A *stationary distribution* of the Markov chain is any distribution $f(x)$ such that

$$f(\theta^{(t)}) = \int p(\theta^{(t)}|\theta^{(t-1)}) f(\theta^{(t-1)}) d\theta^{(t-1)}.$$

That is, if we have a sample from the stationary distribution and update the Markov chain, then the next iterate also follows the stationary distribution. However, not all Markov chains have stationary distributions. The random walk on the integers, for example, does not have a stationary distribution. In this case, the marginal distribution of $\theta^{(t)}$ diverges as t grows.

If we can set up a Markov chain so that it does have a stationary distribution and this stationary distribution equals the target (posterior) distribution, we can use a single draw from the target distribution, $\theta^{(0)}$, to generate a full sample from the target distribution by growing the Markov chain forward in time to $\{\theta^{(0)}, \ldots, \theta^{(N)}\}$.

In practice, however, we cannot typically obtain even one sample from the stationary distribution. Nevertheless, we can consider what a Markov chain at stationarity delivers. The marginal distribution of $\theta^{(t)}$ converges to the stationary distribution as t grows. (Technically, this requires certain regularity conditions, such as an irreducible and aperiodic Markov chain; see Brooks et al. (2011) and references therein. These conditions typically hold in practice.) That is, if we iterate and grow the Markov chain long enough its final values are approximately from the stationary distribution. Thus, after sufficient burn-in, i.e., after the chain appears to reach its stationary distribution, we treat $\{\theta^{(t)}, t = N_0, \ldots N\}$ as a *dependent* sample from the stationary distribution. This is an *approximation*, and it is crucial to diagnose convergence to the stationary distribution as we discuss in Section 4.1. Therefore, we aim to find a Markov chain with stationary distribution equal to the target distribution (i.e., the posterior distribution when doing Bayesian inference). We use this Markov chain to obtain an approximate dependent MC sample from the stationary distribution. This is the basis for MCMC.

3　Markov Chain Monte Carlo

3.1　The Metropolis Sampler

The simplest MCMC algorithm is the *Metropolis sampler*. To explain the Metropolis sampler, we use a generic unknown parameter θ and generic data y. Both θ and y may be univariate or multivariate. (We use this convention when describing the various MCMC methodologies in the remainder of this chapter.) After drawing $\theta^{(0)}$ from some starting distribution, for $t = 1, 2, 3, \ldots$

> Step 1. Sample θ^* from the jumping-rule distribution $J_t(\theta^* | \theta^{(t-1)})$.
>
> Step 2. Compute $r = \frac{p(\theta^* | y)}{p(\theta^{(t-1)} | y)}$.
>
> Step 3. Set $\theta^{(t)} = \begin{cases} \theta^* & \text{with probability } \min(r, 1), \text{ i.e., we } accept \ \theta^*. \\ \theta^{(t-1)} & \text{otherwise, i.e., we } reject \ \theta^*. \end{cases}$

Here, $J_t(\theta^* | \theta^{(t-1)})$ is the *jumping rule*, a distribution depending on the most recent value in the chain that is used to generate a proposal, θ^*, for the next value in the chain. For the Metropolis sampler J_t must be symmetric: that is, $J_t(\theta^* | \theta^{(t-1)}) = J_t(\theta^{(t-1)} | \theta^*)$. Typical choices for $J_t(\theta^* | \theta^{(t-1)})$ when θ is univariate are a uniform distribution centered at $\theta^{(t-1)}$,

$$\text{UNIFORM}(\theta^{(t-1)} - k, \theta^{(t-1)} + k),$$

or a Gaussian distribution centered at $\theta^{(t-1)}$,

$$\text{NORM}(\theta^{(t-1)}, k),$$

where k determines the the size of the "steps" the Markov chain can take while exploring the parameter space. (For the Gaussian jumping rule, k is the variance of the Gaussian distribution.) If θ is multivariate, a common choice of jumping rule is the multivariate Gaussian distribution centered at $\theta^{(t-1)}$,

$$\text{NORM}_d(\theta^{(t-1)}, kI),$$

where $\text{NORM}_d(\vec{\mu}, \Sigma)$ is a d-dimensional multivariate Gaussian distribution with mean vector $\vec{\mu}$ and variance-covariance matrix Σ, I is the identity matrix and k is the common variance term along the diagonal of the variance-covariance matrix given by kI.

Note that "jumps" to higher density are always accepted. That is, if $p(\theta^*|y) > p(\theta^{(t-1)}|y)$, $\frac{p(\theta^*|y)}{p(\theta^{(t-1)})}$ is greater than one and so $\theta^{(t)} = \theta^*$ with probability one. However, jumps to lower density are only sometimes accepted. If $p(\theta^*|y) < p(\theta^{(t-1)}|y)$, then $\theta^{(t)} = \theta^*$ with probability equal to the ratio of the density evaluated at θ^* to the density evaluated at $\theta^{(t-1)}$.

As an example of a Metropolis sampler in practice, we use a simplified model for high-energy spectral analysis. Consider a perfect detector (i.e., constant effective area, no background contamination, perfectly recorded photon energies, etc.) with 1000 energy bins equally spaced from 0.3keV to 7.0keV. A simple power-law model for the observed data is

$$Y_i \sim \text{POISSON}\left(\alpha E_i^{-\beta}\right), \text{ with } \theta = (\alpha, \beta), \tag{3.1}$$

where Y_i are the counts, E_i are the energies, and α and β are model parameters. With prior distribution $p(\alpha, \beta) \propto 1$, the posterior distribution has the form

$$p(\theta|Y) \propto \prod_{i=1}^{n} e^{-[\alpha E_i^{-\beta}]}[\alpha E_i^{-\beta}]^{Y_i}$$

$$= e^{-\alpha \sum_{i=1}^{n} E_i^{-\beta}} \alpha^{\sum_{i=1}^{n} Y_i} \prod_{i=1}^{n} E_i^{-\beta Y_i}.$$

We simulate 2288 counts with $\alpha = 5.0$ and $\beta = 1.69$, and the data are presented in Figure 4. To explore the posterior distribution of the normalization parameter, α, and power-law parameter, β, we use a two-dimensional multivariate Gaussian jumping rule centered at the current sample, $(\alpha^{(t-1)}, \beta^{(t-1)})$, with standard deviations equal to 0.08 and correlation zero:

$$\begin{pmatrix} \alpha^{(t)} \\ \beta^{(t)} \end{pmatrix} \sim \text{NORM}_2 \left(\begin{pmatrix} \alpha^{(t-1)} \\ \beta^{(t-1)} \end{pmatrix}, \begin{pmatrix} 0.08^2 & 0 \\ 0 & 0.08^2 \end{pmatrix} \right). \tag{3.2}$$

One way to visually evaluate the efficiency of an MCMC sampler is to plot the time series or *trace plot* of the samples—the parameter values drawn as a function of their iteration number; see Figure 5. When proposals are rejected, the Markov

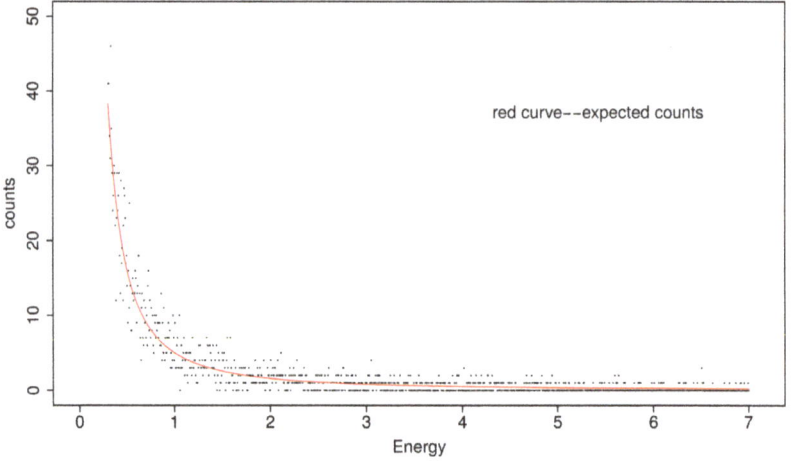

Fig. 4. A simulated data set from the hypothetical perfect detector given in (3.1). The red curve is the expected counts under the model in (3.1), dots are the 2288 simulated counts in the 1000 energy bins.

chain "sticks" at a particular draw. Conversely, if proposals are (almost) never rejected, then the chain typically has high autocorrelation because the current draw is typically close to the previous draw. This can be diagnosed using autocorrelation plots, as in the left column of Figure 6. A high autocorrelation at a lag of 80, for example, means that the draw at iteration t is highly correlated with the draw that came before at iteration $t-80$. Thus autocorrelation measures the *lack* of independence of the Metropolis draws, or how dependent a particular draw is on the history of the chain. Thus we prefer small autocorrelation, and an autocorrelation that decreases sharply as lag increases. For example, in Figure 6 the autocorrelation plot for β is encouraging, while that for α may be a cause for concern.

Two scenarios lead to high autocorrelation: (i) the proposals are (almost) never rejected (i.e., the chain takes small steps), and (ii) the proposals are (almost) never accepted (i.e., the proposed jumps are too far away from the current location). There is thus a tradeoff between acceptance rate—the proportion of proposals that are accepted—and autocorrelation. We discuss strategies for achieving an optimal balance in Section 4.

Once we are satisfied that the Markov chain has produced an MC sample that represents the target posterior distribution well, we can summarize the posterior distribution to perform inference on θ. We first graphically examine the marginal posterior distributions of α and β; see the second column in Figure 6. We also examine their joint posterior distribution, which is presented in Figure 7. The

Time Series Plot for Metropolis Draws

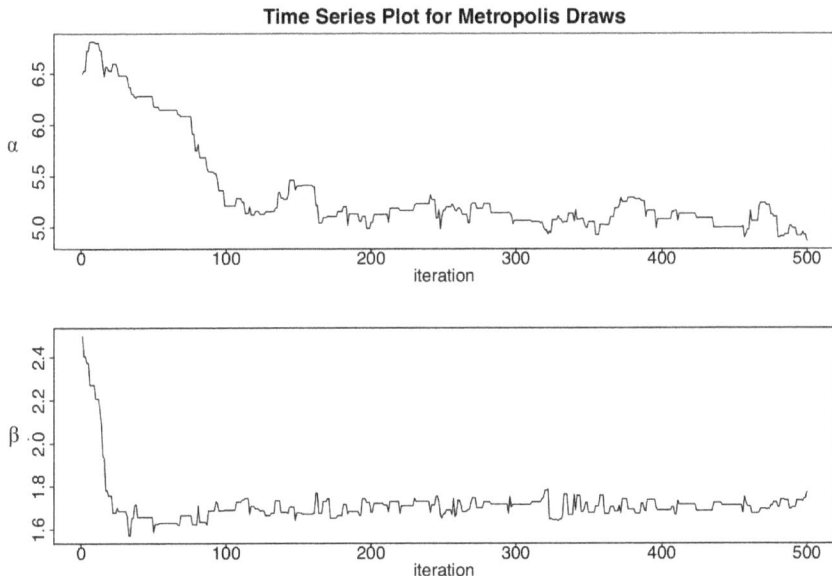

Fig. 5. Using the Metropolis algorithm with jumping rule given in (3.2) to sample α and β for the perfect detector example given in (3.1). The two panels correspond to the trace plot of Metropolis draws for α and β. Chains "stick" at a particular draw when proposals are rejected.

posterior distribution provides a full statistical summary of the unknown model parameters, α and β. Further, point estimates, error bars, and Bayesian credible intervals are all easily calculated using the Metropolis draws. For example, a point estimate for α is estimated by the mean of the Metropolis draws for α, so $E(\alpha|Y) \approx$ 5.13. Because the marginal posterior distributions are roughly symmetrical and bell-shaped, error bars and credible intervals can be computed as the standard deviation of the Metropolis draws and the quantiles of the draws, respectively. With this, $SD(\alpha|Y) \approx 0.11$, and a 95% credible interval is $(4.92, 5.41)$. Similarly, for β we find $E(\beta|Y) \approx 1.71$, $SD(\beta|Y) \approx 0.03$, and a 95% credible interval is $(1.65, 1.76)$.

3.2 The Metropolis-Hastings Sampler

The Metropolis sampler requires the jumping-rule distribution J_t to be symmetric. A generalization of the Metropolis sampler that does not have this restriction is the *Metropolis-Hastings sampler*. It proceeds as follows after drawing $\theta^{(0)}$ from some starting distribution.

For $t = 1, 2, 3, \ldots$

  Step 1. Sample θ^* from $J_t(\theta^*|\theta^{(t-1)})$.

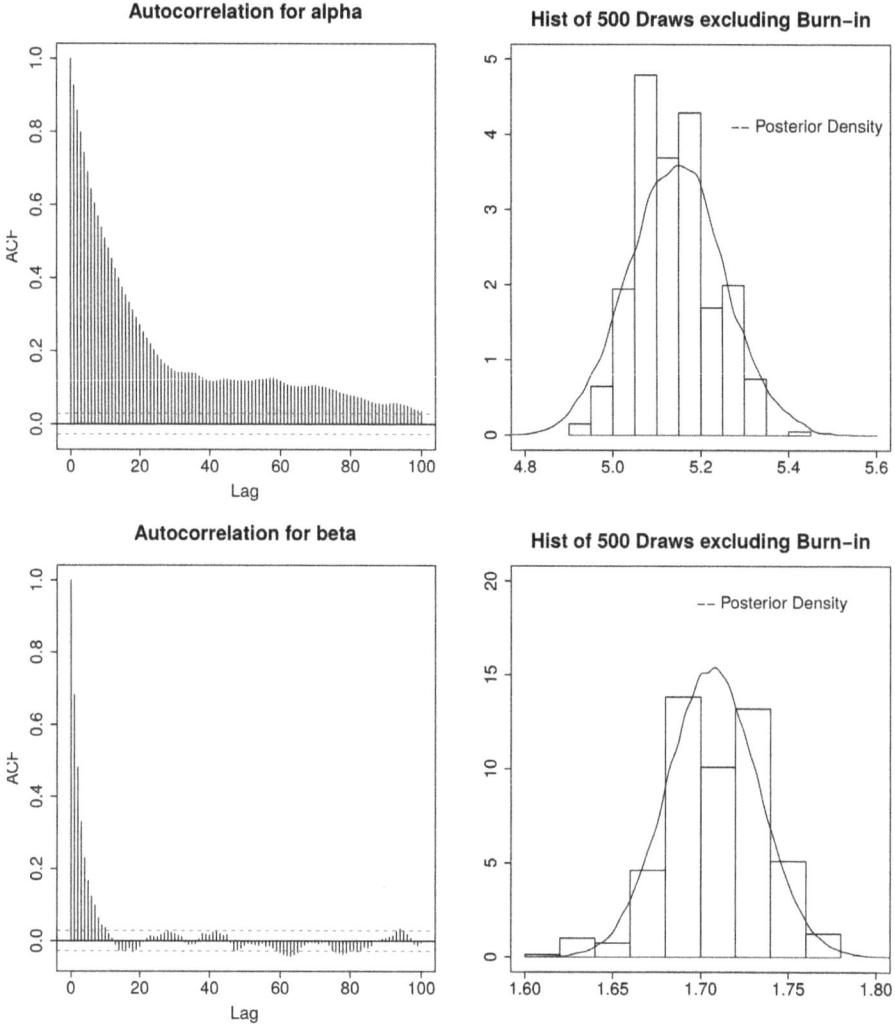

Fig. 6. Autocorrelation plots and the marginal posterior distribution resulting from the Metropolis sampler for the perfect-detector model in (3.1). The autocorrelation plot examines the lack of independence of the Metropolis draws, and it is preferred that autocorrelation drops off quickly as lag increases so that a particular draw is less dependent on the previous history of the chain. Once we are satisfied that the MC sample represents the target posterior distribution well, it can be used to calculate point estimates, error bars, and Bayesian credible intervals as described in Section 2.

Step 2. Compute $r = \dfrac{p(\theta^*|y)/J_t(\theta^*|\theta^{(t-1)})}{p(\theta^{(t-1)}|y)/J_t(\theta^{(t-1)}|\theta^*)}$.

Scatter Plot of Posterior Distribution

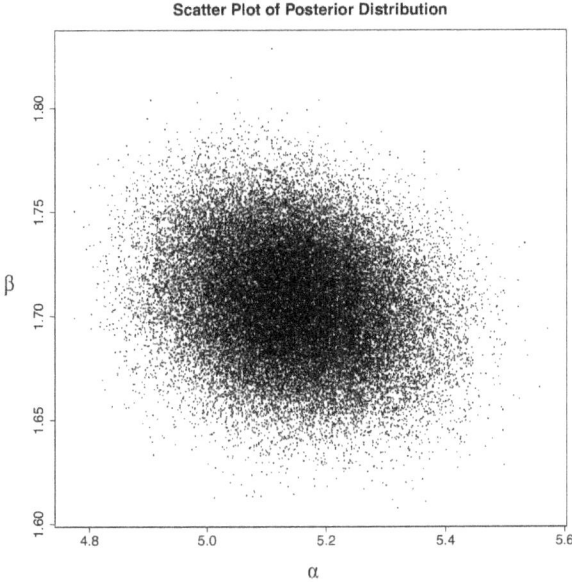

Fig. 7. An MC sample from the joint posterior distribution of α and β under model (3.1). Each dot represents one MC draw obtained with the Metropolis sampler.

$$\text{Step 3. Set } \theta^{(t)} = \begin{cases} \theta^* & \text{with probability } \min(r, 1). \\ \theta^{(t-1)} & \text{otherwise.} \end{cases}$$

Because J_t need not be symmetric and may be any jumping rule, the updated r corrects for bias that would otherwise be present.

A useful Metropolis-Hastings sampler is the *Independence Sampler*, which typically uses an approximation to the posterior as the jumping rule. Using a Gaussian approximation to the posterior distribution, for example,

$$J_t = \text{NORM}_d(\text{MAP estimate, curvature-based estimate of variance matrix}),$$
$$(3.3)$$

where the maximum a posteriori (MAP) estimate is the value of θ that maximizes $p(\theta|y)$ and the curvature-based estimate of the variance matrix is given by

$$\left[-\frac{\partial^2}{\partial\theta \cdot \partial\theta} \log p(\theta|y) \right]^{-1} .$$

Here, $J_t(\theta^*|\theta^{(t-1)})$ is not symmetric and does not even depend on $\theta^{(t-1)}$. To illustrate the independence sampler we return to the perfect-detector model for high-energy spectral analysis, using the same model in (3.1) and simulated data in Figure 4. The model in (3.1) can be identified as a simple *loglinear model*, a special case of a *Generalized Linear Model* (GLM):

$$Y_i \sim \text{Poisson}(\lambda_i) \quad \text{with} \quad \log(\lambda_i) = \log(\alpha) - \beta \log(E_i). \qquad (3.4)$$

Time Series Plot for Metropolis Hastings Draws

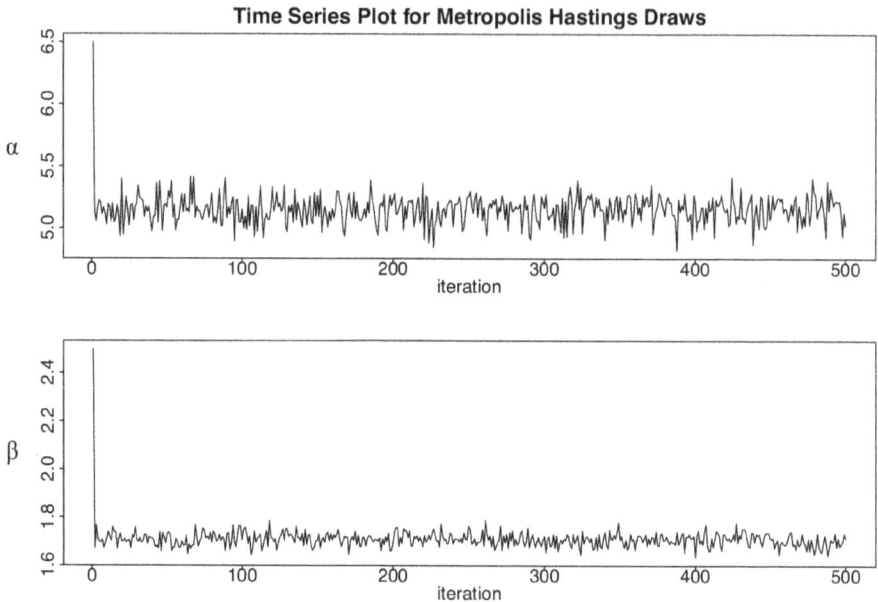

Fig. 8. Trace plots for an independence sampler of the posterior distribution under the model in (3.1). If the proposal distribution approximates the target posterior distribution well, the independence sampler exhibits very little sticking. This is the case in this example.

The model in (3.4) is called a log-linear model because $\log(\lambda)$ is a linear function of $\log(E)$, where $\log(\cdot)$ is the natural logarithm. Ordinary linear regression requires the errors to follow a Gaussian distribution. A GLM extends (i.e., generalizes) ordinary linear regression to allow errors that follow other distributions. Our Poisson model can be fit with the `glm` function in R:

```
> glm.fit = glm( Y~I(-log(E)), family=poisson(link="log") )
> glm.fit$coef     #### MLE of (log(alpha), beta)
> vcov( glm.fit ) #### variance-covariance matrix
```

In this code the MLE of $(\log(\alpha), \beta)$ is stored in `glm.fit$coef` and the variance-covariance matrix is stored in `vcov( glm.fit )`. (Because we are using a uniform prior distribution the posterior distribution is proportional to the likelihood and the MLE equals the MAP estimate.) Alternatively, we can fit (α, β) directly with a general (but perhaps less stable) mode finder, which requires coding the likelihood and specifying starting values.

Using the jumping rule in (3.3) in an independence sampler, the resulting trace plots are presented in Figure 8. Notice that there is very little "sticking" here; the

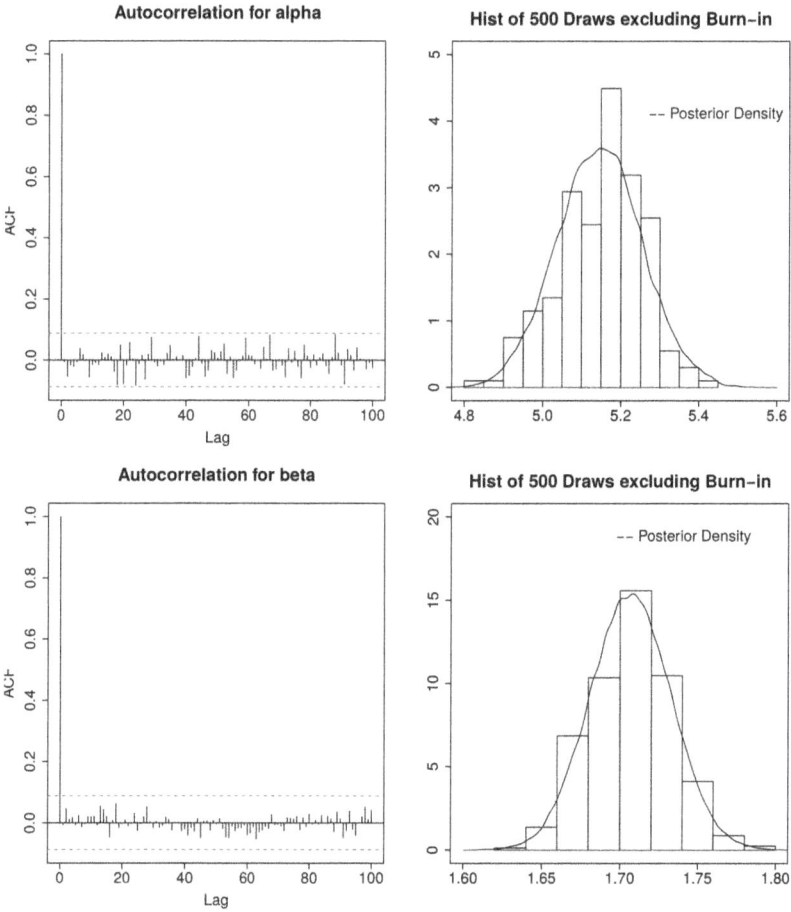

Fig. 9. Autocorrelation plots and the marginal posterior distribution resulting from the independence sampler for the perfect-detector model in (3.1). The autocorrelation plot examines the lack of independence of the Metropolis draws, and it is preferred that autocorrelation drops off quickly as lag increases so that a particular draw is less dependent on the history of the chain. Comparing these plots with those in Figure 6, notice that the autocorrelation is much improved relative to the Metropolis sampler, and the marginal posterior distribution from the two methods agree.

acceptance rate is 98.8%. As with the Metropolis sampler, we can examine the autocorrelation plots and marginal posterior distributions, which are presented in Figure 9. The autocorrelations are nearly zero for all positive lags and, unlike with the Metropolis sampler, because the jumping rule does not depend on $(\alpha^{(t-1)}, \beta^{(t-1)})$ there is no tradeoff between acceptance rate and autocorrelation.

That is, for the independence sampler the acceptance rate is very high and the autocorrelation is very low. (Further, the marginal posterior distributions obtained via the two methods are in agreement, which provides a useful check that the algorithms are working as expected.) However, while it is tempting to therefore always rely on an independence sampler, it is important to acknowledge that this result depends critically on access to a very good approximation to the posterior distribution. Referring back to Figure 7, we see that the joint posterior distribution appears roughly Gaussian, which motivates our choice of jumping rule for the independence sampler. If the joint posterior distribution appeared otherwise (e.g., as in Figure 1), a good approximation may be difficult to construct.

4 Practical Challenges and Advice

4.1 Diagnosing Convergence

As discussed in Section 2.1, after sufficient burn-in, i.e., after a sufficient number of iterations, we treat the MCMC draws as a (correlated) sample from the stationary distribution of the Markov chain. Further, we can construct the Markov chain such that the stationary distribution is the posterior distribution we wish to explore. But how do we determine a sufficient number of burn-in iterations? That is, how do we know that the Markov chain has *converged* sufficiently so that we can regard its iterations as representative of its stationary distribution? Examining the trace plot of a single chain is not sufficient. For example, in Figure 10 we present three trace plots for three Markov chains. The chains have the same target distribution but differ in the supplied starting values. If we saw only the beginning of chain 3, we might falsely conclude that the chain had converged. Similarly, if we only saw chains 1 and 2 we would miss the potential mode that appears to be picked up by chain 3. Therefore, to diagnose convergence it is crucial to compare results from multiple chains that are started from distant points in the parameter space. The chains should be run until they appear to give similar results or they find different solutions (i.e., multiple modes in the posterior distribution).

When using multiple chains (with distant starting values), a quantitative measure of convergence is given by the Gelman and Rubin $\hat{R}$ statistic (Gelman & Rubin, 1992). Consider M chains of length N: $\{\theta_{nm}, n = 1, \ldots, N\}$, where θ_{nm} indicates draw $\theta^{(n)}$ from chain m for $m = 1, \ldots, M$. We can compute the between-chain variance,

$$B = \frac{N}{M-1} \sum_{m-1}^{M} (\bar{\theta}_{\cdot m} - \bar{\theta}_{\cdot\cdot})^2,$$

and the within-chain variance,

$$W = \frac{1}{M} \sum_{m=1}^{M} s_m^2 \ , \ \text{where} \ \ s_m^2 = \frac{1}{N-1} \sum_{n=1}^{N} (\theta_{nm} - \bar{\theta}_{\cdot m})^2,$$

with $\bar{\theta}_{\cdot m}$ being the sample mean of the draws for chain m, and $\bar{\theta}_{\cdot\cdot}$ being the

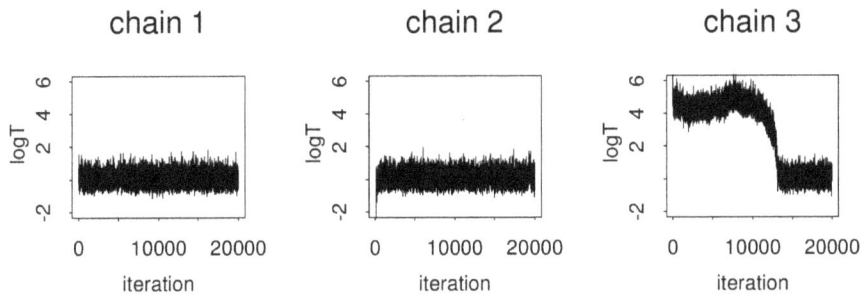

Fig. 10. Multiple chains with different starting values. Relying on only a single chain may lead to a misdiagnosis of convergence.

sample mean of the draws from all the chains combined. There are two ways we can estimate $\mathrm{Var}(\theta \mid Y)$:

1. W, which is an under estimate of $\mathrm{Var}(\theta \mid Y)$ for any finite N.

2. $\widehat{\mathrm{var}}^+(\theta \mid Y) = \frac{N-1}{N}W + \frac{1}{N}B$, which is an over estimate of $\mathrm{Var}(\theta \mid Y)$, so long as the chains have starting values that are distant in the parameter space.

Taking advantage of these two estimates, the $\hat{R}$ statistic is given by

$$\hat{R} = \sqrt{\frac{\widehat{\mathrm{var}}^+(\theta \mid Y)}{W}}.$$

As the chains converge, $\hat{R}$ converges to one (from above). While the above formulas make $\hat{R}$ seem a bit cumbersome to compute, the coda package in R (http://cran. r-project.org/web/packages/coda/index.html) simplifies the task.

4.2 Choosing a Jumping Rule

As stated in Section 3.1, a higher acceptance rate is not always better, particularly for a random-walk Metropolis sampler. In Figure 11 we demonstrate the effect of the jumping rule on the power law parameter from the perfect-detector model in (3.1). The jumping rule is $\mathrm{NORM}(\theta^{(t-1)}, \sigma^2)$, where σ is set to 0.005, 0.08 and 0.4 for the runs that produced the top, middle, and bottom rows, respectively. In the top row, with $\sigma = 0.005$, we observe that the chain takes small steps. This results in a high acceptance rate, 87.5%, but also high autocorrelation. As a result, the chain takes a long time to explore the parameter space and/or converge to its stationary distribution. With $\sigma = 0.08$, in the second row, there is medium autocorrelation and an acceptance rate of 31.6%. Here, the chain takes medium steps and explores the parameter space in an efficient manner. In the third row, with $\sigma = 0.4$, the chain "sticks" for many iterations, yielding a low acceptance rate (3.1%) and high autocorrelation. Here, the chain proposes steps that are often too

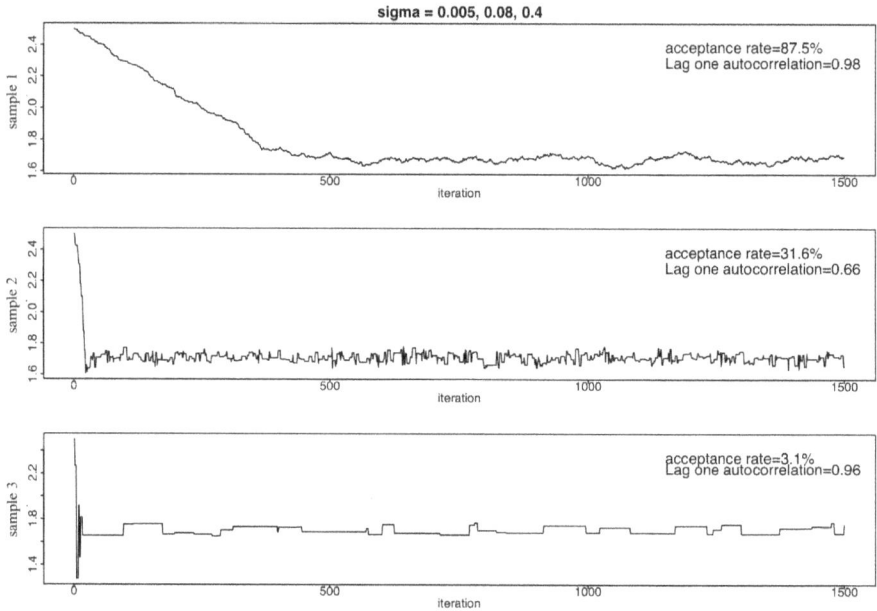

Fig. 11. Effect of the jumping rule on the power law parameter for the perfect-detector model in (3.1). In the top panel the jumping rule is $\text{NORM}(\theta^{(t-1)}, 0.005^2)$ and the chain takes small steps and has a high acceptance rate (87.5%). In the middle panel the jumping rule is $\text{NORM}(\theta^{(t-1)}, 0.08^2)$ and the chain takes medium-sized steps and has an intermediate acceptance rate (31.6%). In the bottom row the jumping rule is $\text{NORM}(\theta^{(t-1)}, 0.4^2)$ and the chain takes large steps and has a low acceptance rate (3.1%). Of the three jumping rules the intermediate rule in the middle panel is best because is has the smallest lag-one autocorrelation.

large and thus rejects many of them. From this we see that an acceptance rate that is either too high or too low results in high autocorrelation via either small steps or many rejected proposals, respectively. A rule of thumb is that an ideal acceptance rate is 40% for single parameters and 20% for multivariate parameters (e.g., Brooks et al., 2011).

Another useful diagnostic for choosing a jumping rule is the *effective sample size* (ESS) it produces. The ESS involves a ratio of variances, and can be approximated by

$$\frac{T}{1 + 2\sum_{k}^{\infty} \rho_k(\theta)},$$

where T is the total number of posterior draws and $\rho_k(\theta)$ is the lag-k autocorrelation of the parameter θ (Jiao & van Dyk, 2015). ESS estimates the size of an independent sample that would contain the same level of information as the dependent sample. The ESS can be easily calculated using the coda package in

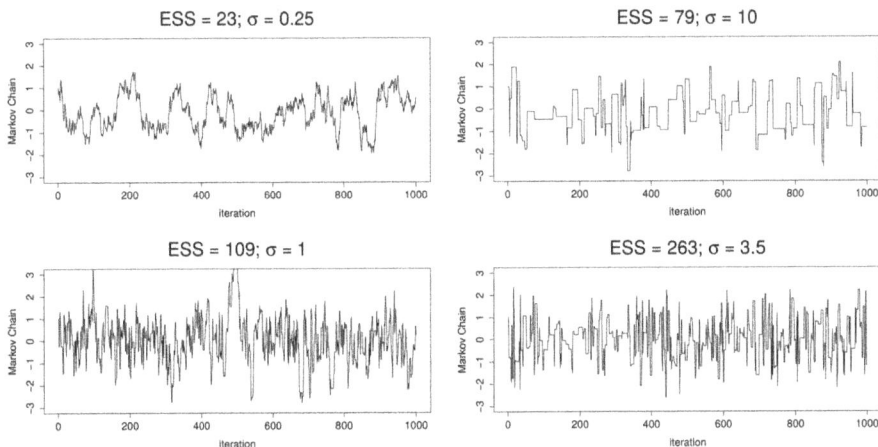

Fig. 12. The effective sample size when sampling from a $\text{NORM}(0, 1)$ target distribution using random-walk Metropolis with a $J_t = \text{NORM}(\theta^{(t-1)}, \sigma^2)$ jumping rule, for $\sigma = 0.25, 1, 3.5,$ and 10. It can be difficult to judge which has the highest ESS from viewing the trace plots alone.

R. If the samples are actually independent, the ESS is equal to the actual sample size.

In general it is not possible to choose an optimal jumping rule by examining trace plots alone. To demonstrate, we sample from a $\text{NORM}(0, 1)$ target distribution using random-walk Metropolis with a $J_t = \text{NORM}(\theta^{(t-1)}, \sigma^2)$ jumping rule, where we set σ equal to 0.25, 1, 3.5 and 10; the results are presented in Figure 12. While we might be able to conclude that the ESS is small when $\sigma = 0.25$ (the chain takes small steps and there is thus high autocorrelation) or when $\sigma = 10$ (the chain sticks often, also inducing high autocorrelation), it is difficult to determine by eye whether $\sigma = 1$ or $\sigma = 3.5$ is preferred. Yet the ESS for $\sigma = 3.5$ is more than double that for $\sigma = 1$. It is therefore important to actually perform the calculation (with the assistance of coda or similar resources)!

If we examine both the ESS and the acceptance rate as a function of $\log(\sigma)$, see Figure 13, we find that high acceptance comes only with small steps, but that the ESS reaches a maximum with an acceptance rate of around 40%, motivating the rules-of-thumb described earlier in this section. However, note that these rules apply to the random-walk Metropolis sampler and are not universal. As we saw previously, the independence sampler is capable of both high acceptance rates and low autocorrelation. (The challenge with the independence sampler is finding a suitable approximation to the target posterior distribution, which may not always be possible in practice.)

Additional challenges arise when θ is multivariate. The tradeoff between autocorrelation and rejection rate is more acute with high posterior correlations

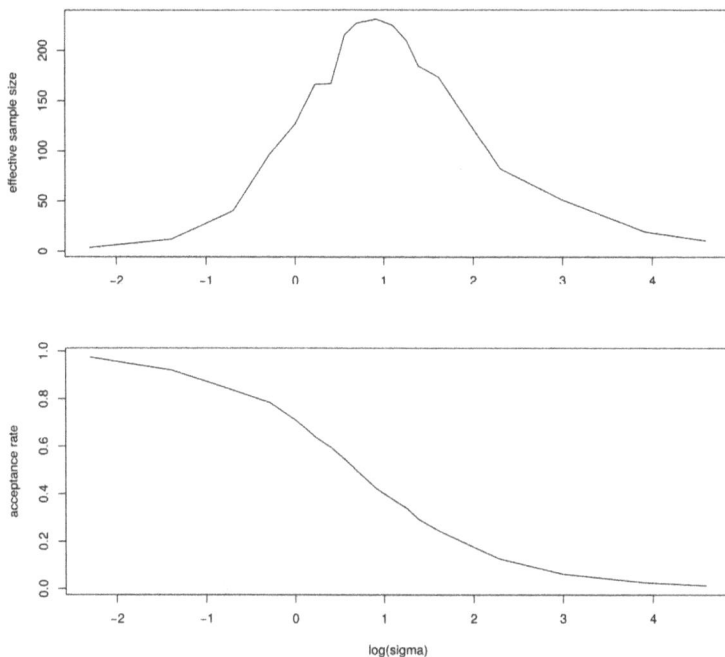

Fig. 13. Finding the optimal acceptance rate when sampling from a NORM$(0, 1)$ target distribution using random-walk Metropolis with a $J_t = \text{NORM}(\theta^{(t-1)}, \sigma^2)$ jumping rule. When the step size, σ, is small, the acceptance rate is high (lower plot). However, because the step size is small the autocorrelation is high, which decreases the ESS (top plot). Conversely, when the step size is high the EES is also low because the chain sticks for many iterations. An optimal acceptance rate, in terms of maximizing ESS, is around 40%. (If this were a multivariate example, we would expect the optimal acceptance rate to be around 20%.)

and high-dimensional parameters. In principle we can use a correlated jumping rule (e.g., a multivariate Gaussian with non-zero terms on the off-diagonal of the variance-covariance matrix), but the desired correlation may vary and it is often difficult to compute in advance. For example, in Figure 14 we plot two highly correlated distributions. For the distribution in the left panel we may be able to construct an efficient jumping rule, though it may take some trial-and-error. However, it is difficult to imagine an efficient random walk jumping rule for the distribution in the right panel; remember that we do not get to see the distribution in advance! Transformations may offer some assistance with both linear and nonlinear correlations, as we demonstrate in Section 4.3.

When constructing a jumping rule it is important to consider the scales of

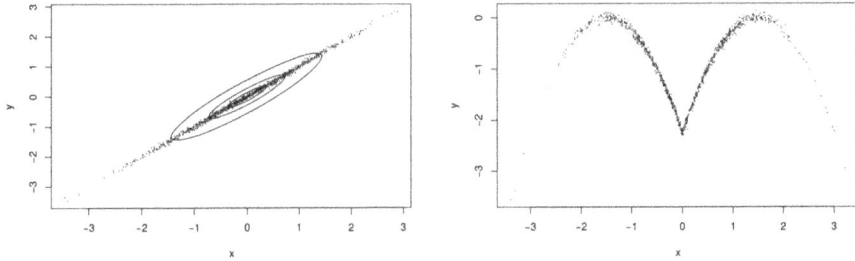

Fig. 14. Two-dimensional distributions with high correlation. While it may be possible to find an efficient random walk jumping rule for the distribution on the left, it is difficult to imagine doing so for the distribution on the right. Remember: we do not get to see the distribution in advance.

the parameters. For example, in Figure 15 the same posterior distribution, that constructed under the model in (3.1), is displayed in the left and right panels. The difference between the panels is the scale on the vertical axes. Because of the difference in scale, α has larger marginal posterior variance compared to β. We can acknowledge this discrepancy and construct a more efficient jumping rule by allowing for the diagonal terms of the variance-covariance matrix (i.e. the variances/standard deviations) for the Gaussian jumping distribution to be different. Of course we do not get to see the joint posterior distribution in advance, but we can use a crude approximation or run a short chain to aid in constructing a more efficient sampler that accounts for the different scales of the parameters. Further, although the posterior correlation between α and β does not appear to be particularly strong, incorporating the correlation into the jumping rule leads to further gains in efficiency.

In Figure 16 we compare the results using a random-walk Metropolis sampler with the original jumping rule (Gaussian with both standard deviations equal to 0.08 and correlation set to zero) to those obtained using a new jumping rule that acknowledges the correlation (Gaussian with standard deviations 0.110 and 0.026 for α and β, respectively, and correlation equal to -0.216). It is clear that convergence is much improved with the new jumping rule; in addition to the decreased autocorrelation, the ESS of the 500 MCMC draws increases from 19 when using the original jumping rule to 75 using the new jumping rule.

General advice for parameters on different scales is to use $\text{NORM}(\theta^{(t)}, kM)$ or $t_\nu(\theta^{(t)}, kM)$, where t_ν is the Student's t-distribution with ν degrees of freedom, location $\theta^{(t)}$, and scale kM, where k is a constant and M is a matrix. The heavier tails of the t-distribution allow for more frequent large jumps compared to using a Gaussian jumping rule, which may result in lower autocorrelation and may reduce the amount of time a chain sticks in the tail of the target distribution or a local mode. For both the Gaussian and t jumping distributions it is advised to use the

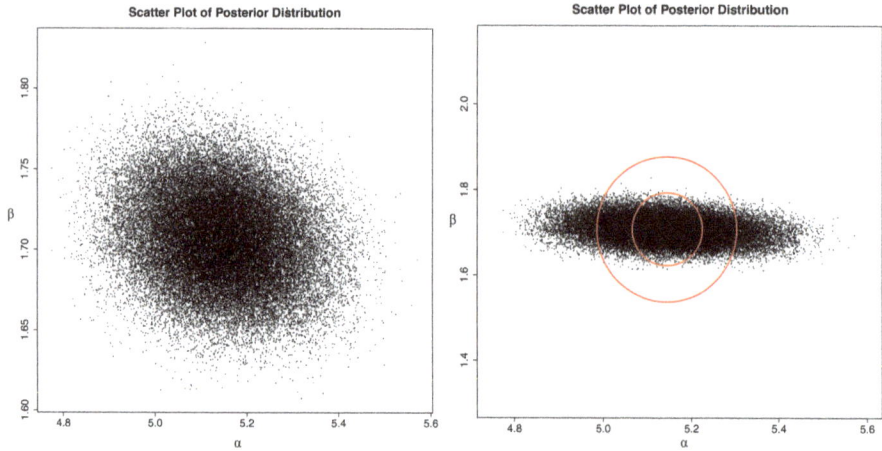

Fig. 15. *Left*: Scatter plot of an MC sample from the posterior distribution under the model in (3.1). *Right*: Scatter plot of the same MC sample after changing the scale to be of equal length on the horizontal and vertical axes. The red circles are one and two standard deviation contours of the jumping-rule distribution in (3.2) when centered at the posterior mode. This Gaussian jumping rule may not be as efficient as one that acknowledges the difference in scale by allowing the diagonal terms of the variance-covariance matrix to be different, i.e., allowing the Gaussian jumping rule to have ellipsoidal contours.

variance-covariance matrix from a standard fitted model for M when such software is available. Adaptive methods are also available which allow the jumping rule to evolve on the fly, though care must be taken with algorithms that use the sampling history in the jumping rule because the chain is no longer a *Markov* chain (see Rosenthal (2011) for a discussion of adaptive samplers). In both approaches, using the variance-covariance matrix from a standard fitted model for M or using an adaptive method to learn and update M on the fly, parameters on different scale are accounted for by allowing the diagonal terms of M to vary. That is, the jumping rule allows for bigger steps for parameters with larger scales. Further, if there are posterior correlations among the model parameters, they are accounted for by allowing non-zero off-diagonal terms in M. It is worth reiterating that one should aim for an acceptance rate of about 20% for a multivariate update or 40% for a univariate update.

4.3 Transformations and Multiple Modes

Transforming the parameters can greatly improve MCMC samplers. For example, in the left panel of Figure 17, we present a density (solid curve) and its normal approximation (dashed curve), i.e., the normal density with the same mean and curvature as the original density; clearly the normal approximation does not agree well with the original density. However, if a square-root transformation, $\sqrt{\theta}$, is

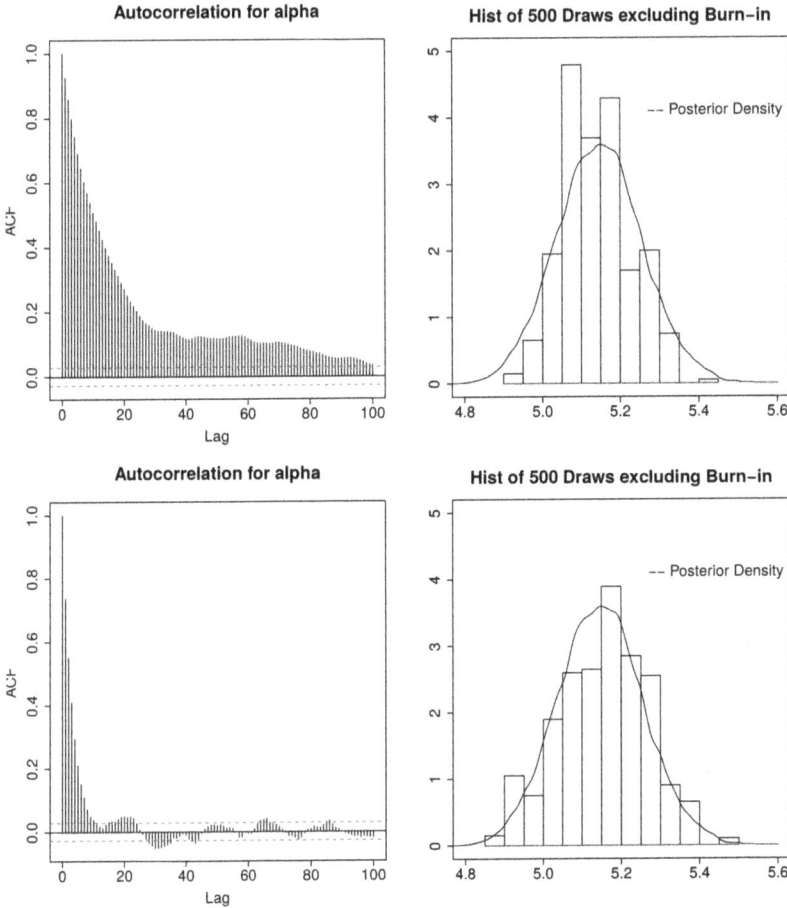

Fig. 16. Incorporating the posterior correlation between α and β under model (3.1) leads to improved convergence when using a Metropolis sampler with Gaussian jumping rule. The original jumping rule is Gaussian with standard deviations equal to 0.08 and correlation set to zero. The new jumping rule is Gaussian with standard deviations 0.110 and 0.026 for α and β, respectively, and correlation equal to -0.216. The original ESS from 500 MCMC draws (corresponding to the results in the top row) is 19, compared to an ESS of 75 for the improved jumping rule.

applied the agreement is much improved; see the right panel of Figure 17.

In general, working with Gaussian or bell-shaped target distributions can lead to more efficient Metropolis and Metropolis-Hastings samplers. Some common strategies to make the target distribution more Gaussian/bell-shaped are to: (i) take the logarithm of positive parameters, (ii) try a square-root transformation

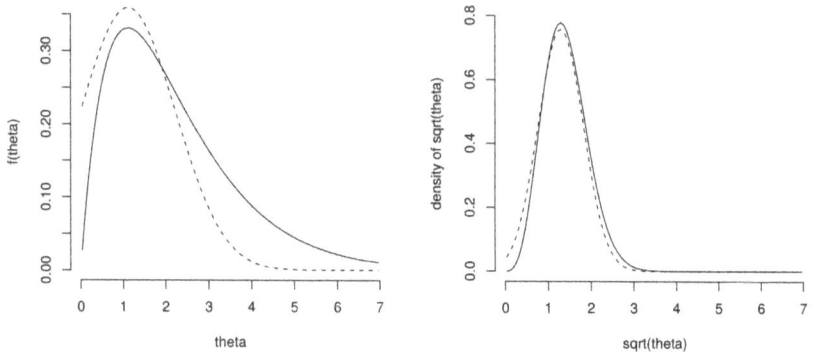

Fig. 17. *Left:* A density for a generic parameter θ (solid curve) and its normal approximation (dashed curve). *Right:* The density of $\sqrt{\theta}$ (solid curve) and its normal approximation (dashed curve) after applying a square-root transformation. There is much better agreement between the density and its normal approximation after the transformation is applied, which can aid MCMC.

if the logarithm is "too strong" (i.e., the logarithm transformation reduces large values more than the square-root transformation, and the latter may be preferred if the resulting transformed density appears more Gaussian), (iii) transform probabilities, p, via the logit transform: $\log(p/(1-p))$, (iv) use more complex transformations for other quantities, and (v) test various transformations using an initial MCMC run.

Strong linear correlations may frustrate MCMC. However, an appropriate choice of linear transformations of θ can remove linear correlations. In Figure 18 we present a distribution of two variables x and y that have high linear correlation (left panel), and the distribution between x and $y - 3x$ (right panel). It is clear that the linear transformation greatly reduces the linear correlation. Linear transformations can also help with non-linear correlations, as illustrated in Figure 19.

A challenge for Bayesian (and frequentist) methods is multiple modes—separate modes of a single target posterior distribution. This means that, for a single target posterior distribution, there are multiple regions of high posterior density that are separated by areas of lower posterior density, as in Figure 20. There may be scientific meaning behind each of the several modes, meaning that multiple parameter combinations may be favored by the current data and prior knowledge, so it is not advised to focus only on the largest mode. One strategy for handling multiple modes is to use a mode finder to "map out" the posterior distribution. Then it is possible to design a jumping rule that accounts for all of the modes, or, perhaps more simply, run separate Markov chains starting at each mode. Another strategy is to use one of several sophisticated computational schemes that are tailored for

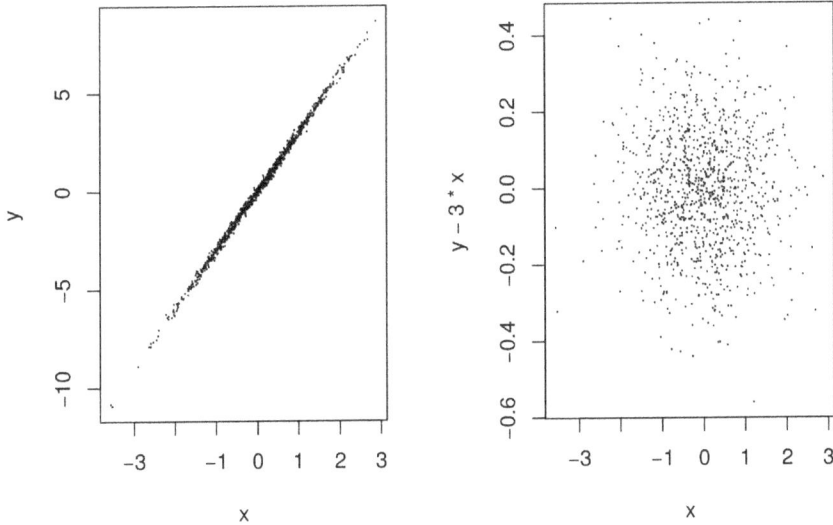

Fig. 18. Linear transformations can help remove linear correlations. In the left panel we plot the distribution between two variables x and y, and in the right panel we plot the distribution of x and a transformed variable $y - 3x$. The distribution in the right panel is clearly more Gaussian, which aids MCMC samplers.

multiple modes. These include (i) adaptive Metropolis-Hastings (e.g., van Dyk & Park, 2011), which adapts the jumping rule when new modes are found, (ii) parallel tempering (Neal, 1996), (iii) nested sampling (Skilling, 2006) and related algorithms (e.g., Feroz et al., 2009), and (iv) the Repelling-Attracting Metropolis (RAM) algorithm (Tak et al., 2017), among others.

4.4   The Gibbs Sampler

In an ideal scenario we would sample directly from the posterior distribution $p(\theta|Y)$ without employing a Metropolis approach. While this may not work in most problems, in *some cases* we can split $\theta = (\theta_1, \theta_2)$, where θ_1 and θ_2 may be univariate or multivariate, so that $p(\theta_1|\theta_2, y)$ and $p(\theta_2|\theta_1, y)$ are both easy to sample although $p(\theta|Y)$ is not. That is, we split the parameter vector θ into blocks such that we can easily sample from the conditional distributions $p(\theta_1|\theta_2, y)$ and $p(\theta_2|\theta_1, y)$ even though sampling from the joint distribution $p(\theta_2, \theta_1|y)$ is challenging. Such an approach is known as *Gibbs Sampling* and proceeds as follows. Starting with some $\theta^{(0)}$, for $t = 1, 2, 3, \ldots$

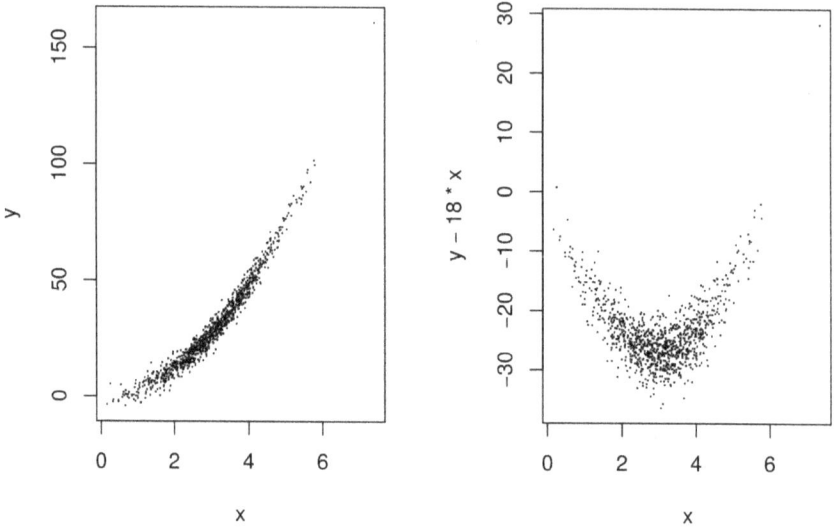

Fig. 19. Linear transformations can also help with non-linear correlations. In the left panel we plot the distribution between two variables x and y; notice that there is strong and non-linear correlation. In the right panel we plot the distribution of x and a transformed variable $y - 18x$. While non-linear correlation remains in the distribution plotted in the right panel, it is not as strong as in the distribution plotted in the left panel.

Step 1.   Draw $\theta_1^{(t)}$ from $p(\theta_1|\theta_2^{(t-1)}, y)$.   (4.1)

Step 2.   Draw $\theta_2^{(t)}$ from $p(\theta_2|\theta_1^{(t)}, y)$.   (4.2)

The Gibbs sampler is a special case of a Metropolis-Hastings sampler in which every proposed jump is accepted, i.e., it can be shown that the acceptance probability is one for every draw.

As an example of a Gibbs sampler in action, recall the perfect-detector model in (3.1):

$$Y_i \sim \text{Poisson}\left(\alpha E_i^{-\beta}\right).$$

Again using $p(\alpha, \beta) \propto 1$ as a prior distribution, recall that the posterior distribu-

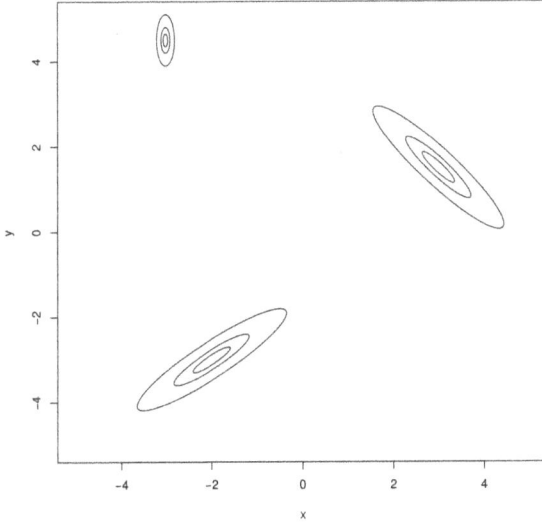

Fig. 20. Contours of a multi-model distribution. This distribution exhibits multiple modes—separated modes of a single target posterior distribution. The multiple modes may each be scientifically meaningful. (Figure adapted from Gelman et al. (2013).)

tion has the form

$$p(\theta|Y) \propto \prod_{i=1}^{n} e^{-[\alpha E_i^{-\beta}]}[\alpha E_i^{-\beta}]^{Y_i}$$

$$= e^{-\alpha \sum_{i=1}^{n} E_i^{-\beta}} \alpha^{\sum_{i=1}^{n} Y_i} \prod_{i=1}^{n} E_i^{-\beta Y_i}.$$

From here we can derive the conditional distribution $p(\alpha|\beta, Y)$:

$$p(\alpha|\beta, Y) \propto e^{-\alpha \sum_{i=1}^{n} E_i^{-\beta}} \alpha^{\sum_{i=1}^{n} Y_i}$$

$$= \text{Gamma}\left(\sum_{i=1}^{n} Y_i + 1, \sum_{i=1}^{n} E_i^{-\beta}\right).$$

However, in this case $p(\beta|\alpha, Y)$ is not a standard distribution:

$$p(\beta|\alpha, Y) \propto e^{-\alpha \sum_{i=1}^{n} E_i^{-\beta}} \prod_{i=1}^{n} E_i^{-\beta Y_i}.$$

While we therefore cannot draw directly from $p(\beta|\alpha, Y)$, we can use a Metropolis or Metropolis-Hastings step to update β within the Gibbs sampler. This resulting sampler is known as a *Metropolis-within-Gibbs Sampler* (e.g., Brooks et al., 2011). An advantage to such a hybrid approach is that Metropolis tends to perform poorly in high dimensions, and Gibbs reduces the dimension. The tradeoff is that Gibbs requires case-by-case probabilistic calculations and may itself be slow to converge. Of course, in practice we always need case-by-case algorithmic development and tuning.

Gibbs sampling tends to work well in cases where there is low to medium posterior correlation, as in Figure 21. There, the autocorrelation with the Gibbs draws is 0.81 and the ESS from 5000 iterations is 525. Gibbs tends to perform poorly in cases where the parameters exhibit high linear posterior correlation or moderate nonlinear correlation. For example, for the distribution in Figure 22, the autocorrelation with the Gibbs draws is 0.998 and the ESS is only 5. Gibbs samplers also perform poorly in the presence of multiple posterior modes. In such cases, the alternative approaches discussed in Section 4.3 should be considered.

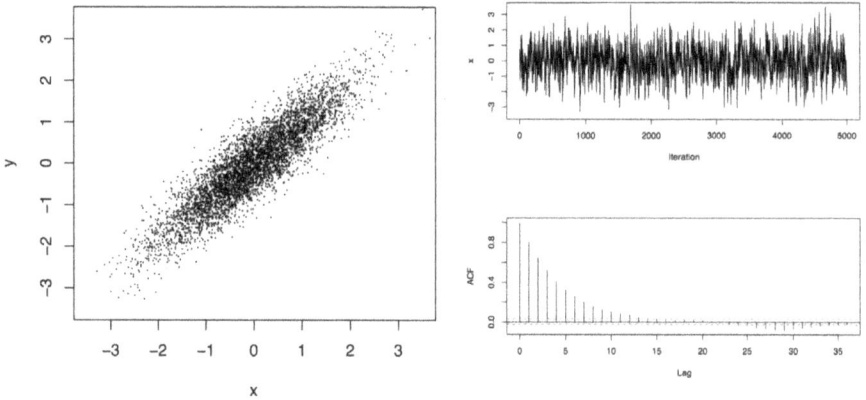

Fig. 21. An example in which the Gibbs sampler performs well in the presence of low to medium posterior correlation. The plot in the left panel is an MC sample of size 5000 from the (target) distribution of two variables x and y. The trace plot of Gibbs draws for x is presented in the top-right panel, and the autocorrelation plot is given in the bottom-right panel. The autocorrelation is 0.81 and the ESS is 525.

The two-step Gibbs sampler given in (4.1)-(4.2) can be generalized to a P-step Gibbs sampler. This general Gibbs sampler breaks θ into P subvectors $\theta = (\theta_1, \ldots, \theta_P)$. The *complete conditional distributions* are given by

$$p(\theta_p|\theta_1, \ldots, \theta_{p-1}, \theta_{p+1}, \ldots, \theta_P, Y), \text{ for } p = 1, \ldots, P.$$

Then, starting with some $\theta^{(0)}$, for $t = 1, 2, 3, \ldots$, the general Gibbs sampler iterates the following steps:

Step 1. Draw $\theta_1^{(t)}$ from $p(\theta_1|\theta_2^{(t-1)}, \ldots, \theta_P^{(t-1)}, Y)$.

$\vdots$

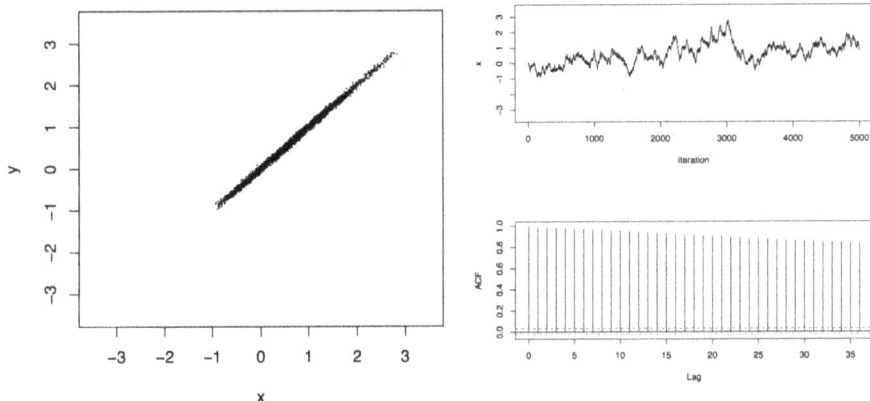

Fig. 22. An example in which the Gibbs sampler performs poorly due to high posterior correlation. The plot in the left panel is an MC sample of size 5000 from the (target) distribution of two variables x and y. The trace plot of Gibbs draws for x is presented in the top-right panel, and the autocorrelation plot is given in the bottom-right panel. The autocorrelation is 0.998 and the ESS is 5. From the autocorrelation plot we can diagnose that the Gibbs sampler has poor efficiency; this is also suggested by the trace plot as successive iterations appear highly correlated.

Step p. Draw $\theta_p^{(t)}$ from $p(\theta_p | \theta_1^{(t)}, \ldots, \theta_{p-1}^{(t)}, \theta_{p+1}^{(t-1)}, \ldots, \theta_P^{(t-1)}, Y)$.

$$\vdots$$

Step P. Draw $\theta_P^{(t)}$ from $p(\theta_P | \theta_1^{(t)}, \ldots, \theta_{P-1}^{(t)}, Y)$.

Note that it is allowable to block subvectors together and should be done if possible. In this way we may update one or several parameters at a time with the aim of increasing efficiency. More general strategies reduce the conditioning on the components of θ in one or more step to improve efficiency and increase the ESS (e.g., the partially collapsed Gibbs sampler; van Dyk & Park, 2008; Park & van Dyk, 2009).

5   Overview of Recommended Strategy

There is an art to constructing an MCMC sampler to obtain a sample from a given target posterior distribution. Our recommended strategy can be broken down into the following steps (adapted from Gelman et al., 2013):

1. Start with a crude approximation to the posterior distribution, perhaps using a mode finder.

2. Use the approximation to set up the jumping rule of an initial sampler (e.g., Gibbs, Metropolis-Hastings, etc.): update one parameter at a time or update parameters in batches.

3. When closed-form complete conditional distributions are available, use Gibbs draws. Otherwise use Metropolis or Metropolis-Hastings jumps.

4. After an initial run, update the jumping rule using the variance-covariance matrix of the initial sample, rescaling so that acceptance rates are near 20% (for multivariate updates) or 40% (for univariate updates).

5. To improve convergence, use transformations so that parameters are approximately independent and/or approximately Gaussian.

6. Run with multiple chains to check for convergence, e.g. using $\hat{R}$.

7. Compare inference based on crude approximation and MCMC. If they are not similar, check for errors before believing the results of the MCMC.

These steps are merely a guide and do not guarantee an efficient sampler. For particularly complex problems researchers may wish to explore alternative MCMC and non-MCMC approaches including, but not limited to, (i) adaptive MCMC (e.g., Rosenthal, 2011), which can automatically improve the jumping rule for more efficient sampling, (ii) Hamiltonian Monte Carlo (Duan et al., 1987; Neal, 2011), which acknowledges the natural geometry of the target (posterior) distribution is particularly useful when there are strong and/or nonlinear posterior correlations, (iii) nested sampling (Skilling, 2006) and related approaches (e.g., Feroz et al., 2009; Handley et al., 2015; Brewer & Foreman-Mackey, 2017), which are particularly useful when faced with multiple modes, and (iv) Approximate Bayesian Computation (e.g. Beaumont et al., 2002; Cameron & Pettitt, 2012; Weyant et al., 2013), which is useful when the likelihood function is unavailable or difficult to evaluate, but forward simulation of synthetic data under the statistical model is relatively fast. Approximate Bayesian Computation in particular is the focus of Chapter 4 in this volume.

References

Beaumont, M.A., Zhang, W., and Balding, D.J. (2002). Approximate Bayesian Computation in Population Genetics. *Genetics*, 162, (4): 2025–2035.

Brewer, B.J. and Foreman-Mackey, D. (2017). DNest4: Diffusive Nested Sampling in C++ and Python. *Journal of Statistical Software*, to appear.

Brooks, S., Gelman, A., Jones, G.L., Meng, X. (eds.) *Handbook of Markov Chain Monte Carlo* (Chapman & Hall/CRC Press, Boca Raton, London, New York, 2011).

Cameron, E. and Pettitt, A.N. (2012). Approximate Bayesian Computation for Astronomical Model Analysis: A Case Study in Galaxy Demographics and Morphological Transformation at High Redshift. *Monthly Notices of the Royal Astronomical Society*, 425, (1): 44–65.

Duane, S., Brian, A.D.K., Pendelton, J. and Roweth, D. (1987). Hybrid Monte Carlo. *Physical Letters B*, 195, (2): 216–222.

Feroz, F., Hobson, M.P., and Bridges, M. (2009). MULTINEST: An Efficient and Robust Bayesian Inference Tool for Cosmology and Particle Physics. *Monthly Notices of the Royal Astronomical Society*, 398, (4): 1601–1614.

Gelman, A., Carlin, J. B., Stern, H. S., Dunson, D. B., Vehtari, A., and Rubin, D. B. *Bayesian Data Analysis: Third Edition* (Chapman & Hall/CRC Press, Boca Raton, London, New York, 2013)

Gelman, A., and Rubin, D.B. (1992). Inference from Iterative Simulation Using Multiple Sequences. *Statistical Science*, 7: 457–511.

Handley, W.J., Hobson, M.P., and Lasenby, A.N. (2015). POLYCHORD: Next-Generation Nested Sampling. *Monthly Notices of the Royal Astronomical Society*, 453, (4): 4384–4398.

Jiao, X. and van Dyk, D.A. (2015). A Corrected and More Efficient Suite of MCMC Samplers for the Multinomial Probit Model. https://arxiv.org/abs/1504.07823.

Neal, R.M. (1996). Sampling from Multimodal Distributions using Tempered Transitions. *Statistics and Computing*, 6, (4): 353–366.

Neal, R.M. (2011). MCMC Using Hamiltonian Dynamics. *Handbook of Markov Chain Monte Carlo* (eds: Brooks, S., Gelman, A., Jones, G.L. and Meng, X.), 113–162.

Park, T. and van Dyk, D. A. (2009). Partially Collapsed Gibbs Samplers: Illustrations and Applications. *Journal of Computational and Graphical Statistics*, 18, 283–305.

Rosenthal, J.S. (2011). Optimal Proposal Distributions and Adaptive MCMC. *Handbook of Markov Chain Monte Carlo* (eds: Brooks, S., Gelman, A., Jones, G.L. and Meng, X.), 93–112.

Skilling, K. (2006). Nested Sampling for General Bayesian Computation. *Bayesian Analysis*, 1, (4): 833–860.

Tak, H., Meng, X., and van Dyk, D.A. (2017). A Repelling-Attracting Metropolis Algorithm for Multimodality. *Journal of Computational and Graphical Statistics*, DOI: 10.1080/10618600.2017.1415911 van Dyk, D. A. and Park, T. (2008). Partially Collapsed Gibbs Samplers: Theory and Methods. *Journal of the American Statistical Association*, 103, 790–796.

van Dyk, D.A. and Park, T. (2011). Partially Collapsed Gibbs Sampling & Path-Adaptive Metropolis-Hastings in High-Energy Astrophysics. *Handbook of Markov Chain Monte Carlo* (eds: Brooks, S., Gelman, A., Jones, G.L. and Meng, X.), 383–400.

Weyant, A., Schafer, C., and Wood-Vasey, W.M. (2013). Likelihood-Free Cosmological Inference with Type Ia Supernovae: Approximate Bayesian Computation for a Complete Treatment of Uncertainty. *The Astrophysical Journal*, 764, (116).

BAYESIAN STATISTICAL METHODS FOR ASTRONOMY
PART III: MODEL BUILDING

David C. Stenning and David A. van Dyk[1]

Abstract. This chapter provides an introduction to Bayesian multi-level models—statistical models with multiple levels of structure with prior distributions specified on all unknown model parameters. Such models have wide applicability in astronomy and astrophysics because information is often available on multiple "levels" that allow complex models to be represented as a sequence of simple sub-models. Hierarchical models are a particular type of multi-level model that describe a population of objects (stars, pixels, etc.) whose parameters follow a common distribution (specified in a lower level of the multi-level model). Bayesian hierarchical models facilitate a concept called "shrinkage," which can produce better estimates of the parameters describing the objects in populations than can simple object-by-object estimators such as arithmetic means. We demonstrate the advantages of using multi-level/hierarchical models and shrinkage estimators via examples from cosmology.

1   Introduction to Multi-Level Models

A key advantage of Bayesian methods is that they allow hierarchical or multi-level structures in data/models, which in turn provide a straightforward mechanism for combining multiple sources of information and/or data streams. An example of a multi-level structure in data/models was given in Chapter 1 Section 1.3, where we introduced the model for background contamination in a single-bin detector. Recall that the detector records contaminated source counts $y = y_S + y_B$, where y_S and y_B are the (unobserved) counts coming from the source and the background, respectively. A background-only exposure is also taken that is 24 times the source exposure, and x background counts are recorded during this exposure. The three-level model is given by

Level 1: $\quad y|y_B, \lambda_S \sim \text{Poisson}(\lambda_S) + y_B$ $\hfill$ (1.1)

---

[1] Statistics Section, Department of Mathematics, Imperial College London

Level 2: $y_B|\lambda_B \sim \text{Poisson}(\lambda_B)$ and $x|\lambda_B \sim \text{Poisson}(\lambda_B \cdot 24)$ (1.2)

Level 3: specify a prior distribution for λ_B, λ_S. (1.3)

Each level of the model specifies a distribution given unobserved quantities whose distributions are given in lower levels. We thus use the multilevel model to break up a complex model for a complex scientific problem into a sequence of relatively simple models. Of course the model in (1)-(3) is relatively simple, but as we illustrate below, multi-level models can be used to describe much more complex physical phenomena. In this way we perform science-driven modeling; each level of the model is interpretable, yet the model can be made increasingly complex and detailed by adding additional levels that are themselves interpretable.

More generally, a *multi-level model* is specified using a series of conditional distributions. With generic random variables X, Y, and Z, and parameter $\theta = (\theta_1, \theta_2, \theta_3)$, the joint distribution can be recovered via the factorization theorem, e.g.,

$$p_{XYZ}(x, y, z|\theta) = p_{X|YZ}(x|y, z, \theta_1) \, p_{Y|Z}(y|z, \theta_2) \, p_Z(z|\theta_3).$$

This model specifies the joint distribution of X, Y, and Z, given the parameter θ. The variables $X, Y,$ and Z may consist of observed data, latent variables, missing data, etc. In this way we can combine models to derive an endless variety of multi-level models.

The rest of this chapter is broken down as follows. In Section 2 we perform a simulation study to demonstrate how to use a multi-level model to account for selection effects. In Section 3 we introduce James-Stein Estimators and the concept of shrinkage. The Bayesian perspective on hierarchical models is covered in Section 4, and we give some concluding remarks and suggestions for further reading in Section 5.

2 A Multilevel Model for Selection Effects

Suppose we wish to estimate the distribution of the absolute magnitudes, M_i, of a population of n objects (e.g., Type Ia supernova), under the model $M_i \sim \text{NORM}(M_0, \sigma^2)$, for $i = 1, \ldots, n$. Suppose that due to selection effects we only observe an object if its apparent magnitude, $m_i = M_i + \mu(z_i)$ is less than 24, where z_i is redshift and $\mu(z_i)$ is a function that arises from the Lambda cold dark matter (Λ-CDM) cosmological model with parameters $\Omega_m = 0.3$, $\Omega_\kappa = 0$, and $H_0 = 67.3$. (Ω_m is the total matter density, Ω_κ is the curvature parameter, and H_0 is the Hubble constant.) Thus M_i and z_i are observed for $N(< n)$ objects, and the aim is to estimate $\theta = (M_0, \sigma^2)$. (For simplicity we assume $\mu(z_i)$ is known for each observed object, which allows us to compute the absolute magnitudes from the apparent magnitudes via $M_i = m_i - \mu(z_i)$. In reality $\mu(z_i)$ depends both on the choice of cosmological model and its unknown parameters.)

We consider two models in turn: a model that ignores the selection effect and a model that takes it into consideration. We then use a simulation study to demonstrate how inference can go awry when important modeling considerations (e.g., the selection effect) are neglected.

2.1 Model 1: Ignore Selection Effect

A naive model that ignores the selection effect specifies a likelihood function that follows from the model

$$M_i|\theta, z_i \overset{\text{indep}}{\sim} \text{NORM}(M_0, \sigma^2), \text{ for } i = 1, \ldots, N.$$

That is, the N observed absolute magnitudes independently follow a Gaussian distribution with mean M_0 and variance σ^2. (The resulting likelihood function does not involve z_i because the selection effect is ignored.) Prior distributions are specified on the unknown model parameters: $M_0 \sim \text{NORM}(\psi, \tau^2)$, and $\sigma^2 \sim \beta^2/\chi_\nu^2$, where $\cdot/\chi^2$ denotes a variable whose reciprocal follows a chi-squared distribution with ν degrees of freedom; multiplying such a variable by β^2 gives a variable distributed as β^2/χ_ν^2. With this model formulation, the complete conditional distributions (see Chapter 2 Section 4.4) of the posterior distribution are :

$$M_0 \mid (M_1, \ldots M_n, \sigma^2) \sim \text{NORM}\left(\bar{M}_0, \ s_{M_0}^2\right),$$

$$\sigma^2 \mid (M_1, \ldots M_n, M_0) \sim \left[\sum_{i=1}^{N}(M_i - M_0)^2 + \beta^2\right]\Big/\chi_{N+\nu}^2,$$

with

$$\bar{M}_0 = \left(\frac{\sum_{i=1}^{N} M_i}{\sigma^2} + \frac{\psi}{\tau^2}\right)\Big/\left(\frac{N}{\sigma^2} + \frac{1}{\tau^2}\right)$$

and

$$s_{M_0}^2 = \left(\frac{N}{\sigma^2} + \frac{1}{\tau^2}\right)^{-1}.$$

If we examine the conditional posterior given σ^2, we notice that the posterior mean, $\bar{M}_0$, is a weighted average of the sample mean ($\frac{1}{N}\sum_{i=1}^{N} M_i$) and prior mean ($\psi$), with weights $\frac{N}{\sigma^2}$ and $\frac{1}{\tau^2}$. Further, as τ^2 goes to infinity, the conditional posterior variance, $s_{M_0}^2$, goes to $\frac{\sigma^2}{N} = \text{Var}\left(\frac{1}{N}\sum_{i=1}^{N} M_i\right)$. That is, we estimate the mean of the (population) distribution of all (i.e. both observed and unobserved) absolute magnitudes as a weighted average of the sample mean of the observed absolute magnitudes and the prior mean of M_0, where the influence of the prior mean depends on the prior variance of M_0 and the number of observed data points. A reference prior for M_0 sets $\psi = 0$ and $\tau^2 = \infty$, which is improper and flat on M_0 (see Chapter 1 Section 2.3 for a discussion of reference priors).

For the conditional posterior given M_0, the prior has the effect of adding ν additional data points with variance β^2. The reference prior for σ^2 sets $\nu = \beta^2 = 0$, which is improper and flat on $\log(\sigma^2)$.

If some set of conditional distributions of the prior and the posterior distribution are of the same family, as they are here, the prior distribution is called that likelihood's *semi-conjugate prior distribution*. See Chapter 1 Section 1.2 for a discussion of conjugate prior distributions. Such priors are very amenable to the Gibbs sampler (Chapter 2 Section 4.4) because the conditional posterior distributions are available in closed form. For this example, to explore the posterior distribution we use the following two-step Gibbs sampler. After drawing $M_0^{(0)}$ and $(\sigma^2)^{(0)}$ from some starting distribution, for $t = 1, 2, 3, \ldots$

Step 1. Update M_0 from its conditional posterior distribution given σ^2:

$$M_0^{(t+1)} \sim \text{NORM}\left(\bar{M}_0, \ s_{M_0}^2\right),$$

$$\text{with} \qquad \bar{M}_0 = \left(\frac{\sum_{i=1}^N M_i}{(\sigma^2)^{(t)}} + \frac{\psi}{\tau^2}\right) \Big/ \left(\frac{N}{(\sigma^2)^{(t)}} + \frac{1}{\tau^2}\right),$$

$$\text{and} \qquad s_{M_0}^2 = \left(\frac{N}{(\sigma^2)^{(t)}} + \frac{1}{\tau^2}\right)^{-1}.$$

Step 2. Update σ^2 from its conditional posterior distribution given μ:

$$(\sigma^2)^{(t+1)} \sim \left[\sum_{i=1}^N \left(M_i - M_0^{(t+1)}\right)^2 + \beta^2\right] \Big/ \chi_{N+\nu}^2.$$

2.2   Model 2: Account for Selection Effect

To account for the selection effect, we introduce an *indicator variable*, O_i, such that $O_i = 1$ if object i is observed, i.e., if $M_i < 24 - \mu(z_i)$, and $O_i = 0$ otherwise. With this, $\Pr(O_i = 1 | M_i, z_i, \theta) = \text{Indicator}\{M_i < 24 - \mu(z_i)\}$, where $\text{Indicator}\{\cdot\}$ is the *indicator function*, which is equal to one if the condition inside the curley brackets is true and is equal to zero otherwise. That is, $\Pr(O_i = 1 | M_i, z_i, \theta) = 1$ if $M_i < 24 - \mu(z_i)$, else $\Pr(O_i = 1 | M_i, z_i, \theta) = 0$; this specific probability function only takes values 0 or 1 and does not take other values on the unit interval. Using Bayes' Theorem (Chapter 1 Section 1.1), the distribution of the observed magnitudes is

$$p(M_i | O_i = 1, \theta, z_i) = \frac{\Pr(O_i = 1 | M_i, z_i, \theta) p(M_i | \theta, z_i)}{\int \Pr(O_i = 1 | M_i, z_i, \theta) p(M_i | \theta, z_i) \mathrm{d} M_i}, \qquad (2.1)$$

with $M_i | \theta, z_i \overset{\text{indep}}{\sim} \text{NORM}(M_0, \sigma^2)$. The right-hand side of (4) simplifies to

$$M_i | (O_i = 1, \theta, z_i) \overset{\text{indep}}{\sim} \text{TRUNNORM}[M_0, \sigma^2; 24 - \mu(z_i)],$$

where $\text{TRUNNORM}[M_0, \sigma^2; 24 - \mu(z_i)]$ is the *truncated normal distribution* with upper truncation $24 - \mu(z_i)$.

If a generic random variable X follows a $\text{NORM}(\delta, \gamma^2)$ distribution but only within the interval (a, b) and cannot take values outside this interval, then X is said to follow a $\text{TRUNNORM}[\delta, \gamma^2; a, b]$ distribution with lower truncation a and upper truncation b. Its probability density function, for $a \leq x \leq b$ is given by

$$p(\delta, \gamma^2; a, b) = \frac{\phi\left(\frac{x-\delta}{\gamma}\right)}{\gamma\left(\Phi\left(\frac{b-\delta}{\gamma}\right) - \Phi\left(\frac{a-\delta}{\gamma}\right)\right)}, \tag{2.2}$$

and zero elsewhere, where $\phi(\cdot)$ is the probability density function of the standard normal distribution (i.e., the $\text{NORM}(0,1)$ distribution), and $\Phi(\cdot)$ is its cumulative distribution function. For the example at hand there is no lower truncation, i.e., $a = -\infty$ in the above formulation for the truncated normal distribution; the upper truncation is at $b = 24 - \mu(z_i)$. Then (5) simplifies to

$$p(M_0, \sigma^2; a = -\infty, b = 24 - \mu(z_i)) = \frac{\phi\left(\frac{x-M_0}{\sigma}\right)}{\sigma\left(\Phi\left(\frac{24-\mu(z_i)-M_0}{\sigma}\right)\right)}. \tag{2.3}$$

We adopt the same prior distributions on the unknown model parameters as for Model 1: $M_0 \sim \text{NORM}(\psi, \tau^2)$, and $\sigma^2 \sim \beta^2/\chi_\nu^2$. However, here the prior is not (semi-)conjugate, and the posterior is not a standard distribution. Thus neither step of the Gibbs sampler is a standard distribution, and so we use a Metropolis-within-Gibbs scheme instead (Chapter 2 Section 4.4). After drawing $M_0^{(0)}$ and $(\sigma^2)^{(0)}$ from some starting distribution, for $t = 1, 2, 3, \ldots$

Step 1. Update M_0 from its conditional distribution given σ^2 using random-walk Metropolis with a $\text{NORM}(M_0^{(t)}, s_1^2)$ proposal distribution.

Step 2. Update σ^2 from its conditional distribution given M_0 using random-walk Metropolis-Hastings with a $\text{LOGNORM}\left[\log\left(\sigma^{2\,(t)}\right), s_2^2\right]$ proposal distribution, where $\text{LOGNORM}(\cdot, \cdot)$ is the *log-normal distribution*. (If the generic random variable X follows a $\text{LOGNORM}(\delta, \gamma^2)$ distribution, then $\log(X)$ follows a $\text{NORM}(\delta, \gamma^2)$ distribution.)

In the above steps, s_1^2 and s_2^2 can be adjusted to obtain an acceptance rate of around 40% using the strategies described in Chapter 2 Section 4.

2.3 Simulation Study

2.3.1 Simulation Study 1

To evaluate the bias caused by ignoring selection effects, we perform a simulation study. First, we generate a simulated data set by sampling the absolute magnitudes: $M_i \sim \text{NORM}(M_0 = -19.3, \sigma = 1)$ for $i = 1, \ldots, 200$ objects. Then we sample the associated redshifts, z_i, from $p(z) \propto (1 + z)^2$. Applying the selection effect yields $N = 112$ observed absolute magnitudes, as illustrated in Figure 1. There, absolute magnitudes are plotted as a function of redshift, with those observed in blue and those missing due to the selection effect in red. The cutoff due

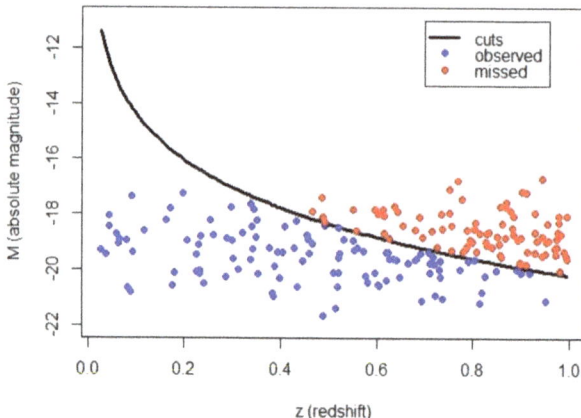

Fig. 1. The data for Simulation Study 1. Blue points are observed absolute magnitudes and red points are missing absolute magnitudes. The selection effect cutoff that causes some absolute magnitudes to be missing is illustrated by the black curve.

to the selection effect, corresponding to $M_i < 24 - \mu(z_i)$, is given by the solid black curve.

The marginal posterior distributions of M_0 and $\log(\sigma^2)$ corresponding to Model 1 (ignoring selection effect) and Model 2 (accounting for selection effect) are given in the two columns of Figure 2. The vertical red lines are the true parameter values, and the histograms are from MCMC draws using the Gibbs samplers described in Sections 2.1 and 2.2. The hyperparameters used to set the prior distributions are $\psi = -19.3$, $\tau = 20$, $\nu = 0.02$, and $\beta^2 = 0.02$. Notice that the inference we obtain from Model 1 nearly misses the true parameter values; the true values, particularly that for M_0, are in the tails of the posteriors. By contrast, accounting for selection effects via Model 2 improves our inferences.

2.3.2   Simulation Study 2

To further illustrate the danger of not including the selection effect in the statistical model, we change the simulation study by increasing the variance in the distribution from which we simulate the data, i.e. the variance of the absolute magnitudes. Specifically, we sample $M_i \sim \text{NORM}(M_0 = -19.3, \sigma = 3)$ for $i = 1, \ldots, 200$. Sampling z_i from $p(z) \propto (1 + z)^2$ and applying the selection effect as done previously yields $N = 101$ observed absolute magnitudes; see Figure 3.

Using the same prior distributions as in Simulation Study 1, we rerun the Gibbs samplers and obtain the results presented in Figure 4. Here, the consequence of ignoring the selection effect is stark: the marginal posterior distributions we obtain with Model 1 completely miss the true parameter values. The marginal posteriors resulting from Model 2, however, cover the true values. Also notice that the

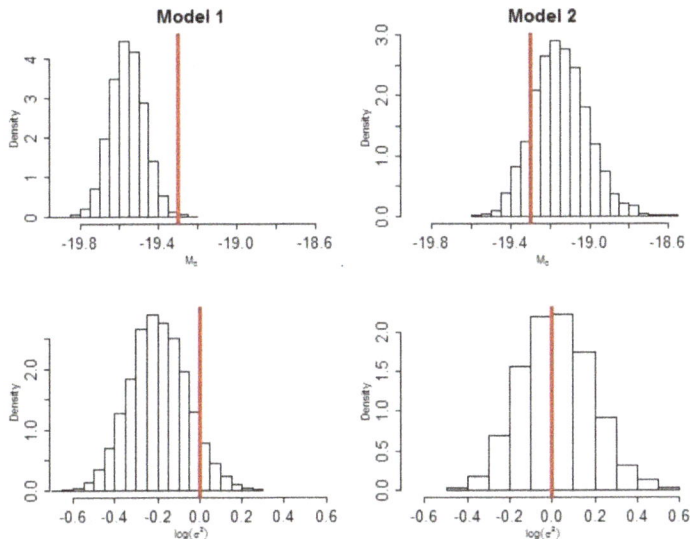

Fig. 2. The results of Simulation Study 1. The red vertical lines are true parameter values, and histograms depict an MC sample from the posterior distribution obtained with MCMC. Notice that Model 2, which accounts for the selection effect, performs better than Model 1, which does not account for the selection effect.

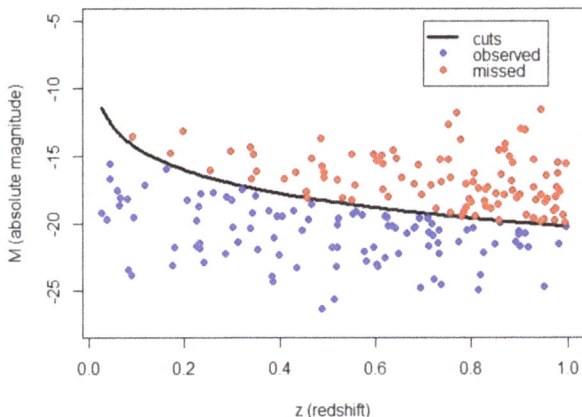

Fig. 3. The data for Simulation Study 2. Blue points are observed absolute magnitudes and red points are missing absolute magnitudes. The selection effect cutoff that causes some absolute magnitudes to be missing is illustrated by the black curve.

marginal posteriors resulting from Model 2 are wider, i.e., the posterior variances are larger. Keep in mind that for a real scientific analysis we do not know the

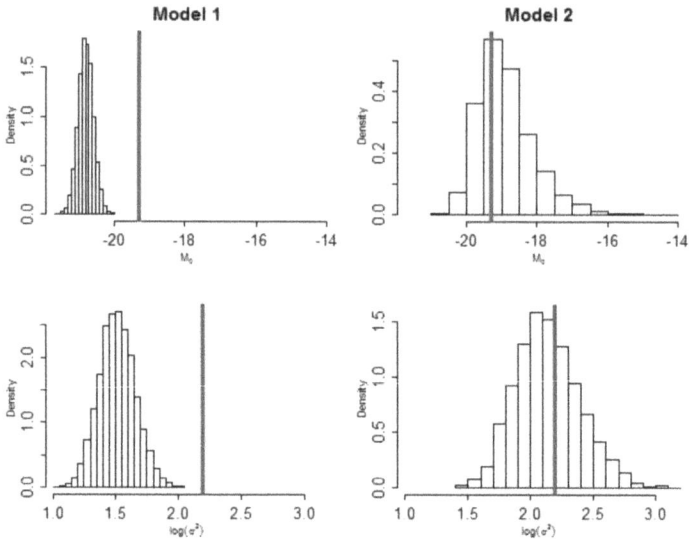

Fig. 4. The results of Simulation Study 2. The red vertical lines are true parameter values, and histograms depict an MC sample from the posterior distribution obtained with MCMC. Notice that the marginal posterior distributions resulting from Model 1, which does not account for the selection effect, fail to cover the true parameter values.

true parameter values. It would be folly to believe that the inferences resulting from Model 1 are preferred simply because they exhibit greater precision. The inferences resulting from Model 2 are less precise but also less biased, and the true values fall within one posterior standard deviation of the posterior means/medians.

3   James-Stein Estimators and Shrinkage

Suppose we wish to estimate a generic parameter, θ, from repeated measurements of some data y:

$$y_i \overset{\text{indep}}{\sim} \text{NORM}(\theta, \sigma^2) \quad \text{for} \quad i = 1, \dots, n.$$

For example, we might take n measures of a known source to calibrate a detector. An obvious estimator of θ is the arithmetic average:

$$\hat{\theta}^{\text{naive}} = \frac{1}{n} \sum_{i=1}^{n} y_i.$$

In evaluating whether this is a "good" estimator, we can consider how far off $\hat{\theta}^{\text{naive}}$ is from θ:

$$(\hat{\theta} - \theta)^2,$$

which is the *squared error*. We can also consider the *mean square error* (MSE) of $\hat{\theta}$, which tells us how far off we expect it to be from θ:

$$\text{MSE}(\hat{\theta} \mid \theta) = \text{E}\left[(\hat{\theta} - \theta)^2 \mid \theta\right] = \int \left(\hat{\theta}(y) - \theta\right)^2 f_Y(y \mid \theta) dy.$$

An estimator is said to be *inadmissible* if there is an estimator that is uniformly better in terms of MSE (or similar measure of discrepancy):

$$\text{MSE}(\hat{\theta} \mid \theta) < \text{MSE}(\hat{\theta}^{\text{naive}} \mid \theta) \quad \text{for all} \quad \theta.$$

The MSE is the average squared error of an estimator over repeated sampling of the data, y.

As an example, suppose π is the true hardness ratio of a spectrum, i.e., π is the expected proportion of hard (high-energy) photons. With a sample of n counts, let H be the observed proportion of hard counts among the n observed counts. Taking $n = 3$, we can model $H \sim \text{BINOMIAL}(n = 3, \pi)$ and recall from Chapter 1 Section 2.1 that if $H|n, \pi \sim \text{BINOMIAL}(n, \pi)$ and $\pi \sim \text{BETA}(\alpha, \beta)$, then the posterior distribution of the true (or large-sample) hardness ratios is $\pi|H, n \sim \text{BETA}(h + \alpha, n - h + \beta)$. Consider four estimates of π:

(i) $\hat{\pi}_1 = H/n$, the maximum likelihood estimator of π

(ii) $\hat{\pi}_2 = \dfrac{H + 1}{n + 2}$; this is $\text{E}(\pi \mid H)$ when π has prior distribution $\pi \sim \text{Beta}(1, 1)$

(iii) $\hat{\pi}_3 = \dfrac{H + 1}{n + 5}$; this is $\text{E}(\pi \mid H)$ when π has prior distribution $\pi \sim \text{Beta}(1, 4)$

(iv) $\hat{\pi}_4 = \dfrac{H + 4}{n + 5}$; this is $\text{E}(\pi \mid H)$ when π has prior distribution $\pi \sim \text{Beta}(4, 1)$.

If the data is a random sample of all possible data sets, the estimator $\hat{\pi}_i$ is also random. It thus has a distribution, a mean and a variance, and we can evaluate the $\hat{\pi}_i$ as an estimator of π in terms of its:

1. bias: $\text{E}(\hat{\pi}_i \mid \pi) - \pi$,

2. variance: $\text{E}\left[\left(\hat{\pi}_i - \text{E}(\hat{\pi}_i \mid \pi)\right)^2 \mid \pi\right]$, and

3. MSE: $\text{E}\left[(\hat{\pi}_i - \pi)^2 \mid \pi\right] = \text{bias}^2 + \text{variance}$.

Bias typically has a bad connotation, but we can deduce from the above that trading a bit of bias for a reduction in variance can lead to a lower MSE. That is, a biased estimator may tend to exhibit smaller errors than an unbiased estimator. Conversely, high precision (i.e., low variance) is not ideal in terms of MSE if there is also high bias, as we discussed in Section 2.3. There is thus a tradeoff between minimizing bias and minimizing variance, appropriately referred to as the *bias-variance tradeoff*. When evaluating an estimator based on its MSE it is

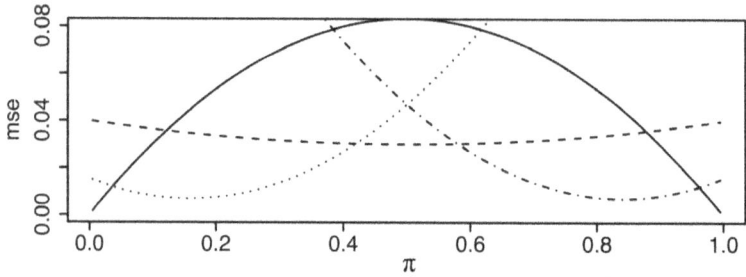

Fig. 5. The MSE for all four estimators of the beta-binomial example as a function of π. The solid curve is the MSE of $\hat{\pi}_1$, the MLE of π, as a function of π. The dotted, dashed, and dot-dashed curves are the MSE of $E(\pi \mid H)$ as a function of π when π has prior distribution Beta$(1,1)$, Beta$(1,4)$, and Beta$(4,1)$, respectively. Because no one estimator is preferred for all values of π, there is no evidence of inadmissibility among these four estimators.

therefore best to seek a balance between bias and variance. For this beta-binomial example, the MSE of all four estimators depends on the true π, as illustrated in Figure 5. Because no one estimator is preferred for *all* values of π, there is no evidence of inadmissibility. (Of course there are many other possible estimators that in principle may dominate one or more of those plotted in Figure 5 in terms of MSE.) Among the four estimators plotted here, $\hat{\pi}_1 = H/n$ is best if estimating a proportion very near zero or one. When estimating a hardness ratio (or other proportion) nearer 0.5, $\hat{\pi}_2$ is better.

For another example, suppose we wish to estimate each of the mean parameters $\theta_1, \ldots, \theta_G$ in the model

$$y_{ij} \overset{\text{indep}}{\sim} \text{NORM}(\theta_j, \sigma^2) \text{ for } i = 1, \ldots, n \text{ and } j = 1, \ldots, G.$$

The obvious estimators, the arithmetic means,

$$\hat{\theta}_j^{\text{naive}} = \frac{1}{n} \sum_{i=1}^{n} y_{ij}, \tag{3.1}$$

are inadmissible if $G \geq 3$. This is because the *James-Stein Estimator*,

$$\hat{\theta}_j^{\text{JS}} = \left(1 - \omega^{\text{JS}}\right) \hat{\theta}_j^{\text{naive}} + \omega^{\text{JS}} \nu , \tag{3.2}$$

with $\omega^{\text{JS}} \approx \dfrac{\sigma^2/n}{\sigma^2/n + \tau_\nu^2}$ and $\tau_\nu^2 = E[(\theta_i - \nu)^2]$, dominates θ^{naive} for any ν.[1]

---

[1] More specifically,

$$\omega^{\text{JS}} = \frac{(G-2)\sigma^2}{n \sum_{j=1}^{G} (\hat{\theta}_j^{\text{naive}} - \nu)^2}.$$

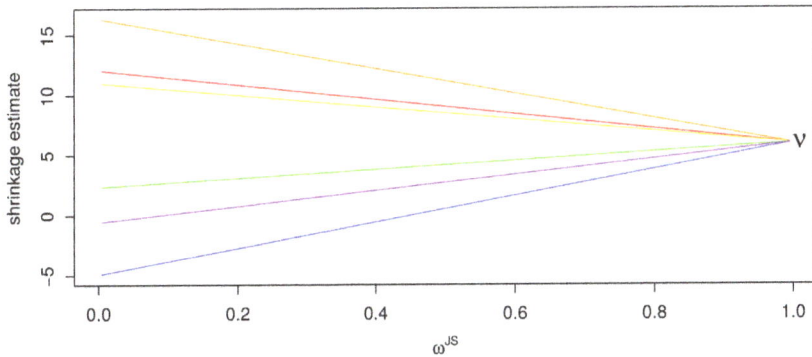

Fig. 6. Illustration of shrinkage with the James-Stein Estimator. Different colors represent parameter estimates for each of six parameters as a function of ω^{JS}. When $\omega^{\mathrm{JS}} = 1$, all estimates are shrunk to the value chosen for ν. This corresponds to complete shrinkage. When $\omega^{\mathrm{JS}} = 0$, parameters are estimated by their sample average. This corresponds to no shrinkage.

The James-Stein Estimator is an example of a *shrinkage estimator*, meaning that the estimates of the individual θ_j are shrunk towards a quantity ν, where the amount of shrinkage is determined by ω^{JS}. This is illustrated in Figure 6. There, we see that when $\omega^{\mathrm{JS}} = 1$, $\hat{\theta}_j^{\mathrm{JS}} = \nu$ for all j. Similarly, when $\omega^{\mathrm{JS}} = 0$, $\hat{\theta}_j^{\mathrm{JS}} = \hat{\theta}_j^{\mathrm{naive}}$ and there is no shrinkage. A good choice of ν in terms of MSE is a value near the center of the distribution of the θ_j. Nevertheless, James-Stein Estimators dominate the naive case-by-case sample averages in (3.1) for *any choice* of ν. (Shrinkage is mild and $\hat{\theta}^{\mathrm{JS}} \approx \hat{\theta}^{\mathrm{naive}}$ for most ν.)

As an illustration, suppose $y_j \sim \mathrm{NORM}(\theta_j, 1)$ for $j = 1, \dots, 10$, with the θ_j evenly distributed on $[0,1]$. The MSE for the naive estimator, $\hat{\theta}^{\mathrm{naive}}$, and that for the James-Stein Estimator, $\hat{\theta}^{\mathrm{JS}}$, for a range of values for ν are given by the blue and red curves in Figure 7, respectively. If instead the θ_j are evenly distributed on $[-4,5]$, the MSE for the two estimators as a function of ν is given in Figure 8. In the latter, because the θ_j are (evenly) distributed over a larger range and thus have larger variance, the difference in MSE between the two estimators is less pronounced for all values of ν; larger variance for the θ_j typically results in less shrinkage. Regardless, such result is astonishing; the *biased* James-Stein estimator (which is biased due to shrinkage) has lower MSE than does the simplest of all estimators, the arithmetic mean! James-Stein represents a milestone in statistical thinking as the results were viewed as paradoxical and counterintuitive when first introduced.

In summary, if estimating more than two parameters it is always better to use

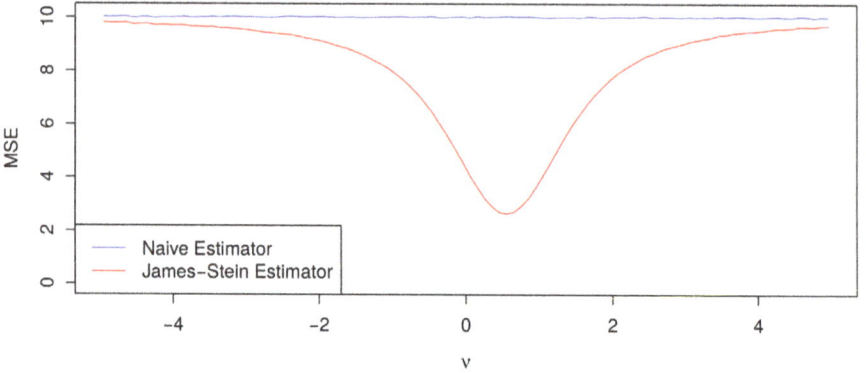

Fig. 7. MSE for the naive estimator (blue) and James-Stein Estimator (red) when the θ_j are evenly distributed on [0,1]. The James-Stein Estimator dominates the naive estimator regardless of the choice of shrinkage term ν. The difference between the two estimators is pronounced because the variance of the θ_j is small.

shrinkage estimators and to optimally shrink toward the average of the parameters. The most gain is to be had when the naive (non-shrinkage) estimators are noisy (σ^2 is large) and/or the set of parameter values are similar (τ_ν^2 is small). From a Bayesian point of view this offers an opportunity, as we explore in the following section.

4   Hierarchical Models and the Bayesian Perspective

From a Bayesian point of view it is quite natural to consider the distribution of a parameter or the *common distribution of a group of parameters*. Models that are formulated in terms of the latter are called *hierarchical models*. For an example of a simple Bayesian hierarchical model, suppose

$$y_{ij}|\theta_j \overset{\text{indep}}{\sim} \text{NORM}(\theta_j, \sigma^2) \quad \text{for} \quad i = 1, \ldots, n \quad \text{and} \quad j = 1, \ldots, G$$

with

$$\theta_j \overset{\text{indep}}{\sim} \text{NORM}(\mu, \tau^2), \tag{4.1}$$

and let $\phi = (\sigma^2, \tau^2, \mu)$. Then,

$$\text{E}(\theta_j \mid Y, \phi) = (1 - \omega^{\text{HB}})\hat{\theta}^{\text{naive}} + \omega^{\text{HB}}\mu \tag{4.2}$$

with

$$\omega^{\text{HB}} = \frac{\sigma^2/n}{\sigma^2/n + \tau^2}.$$

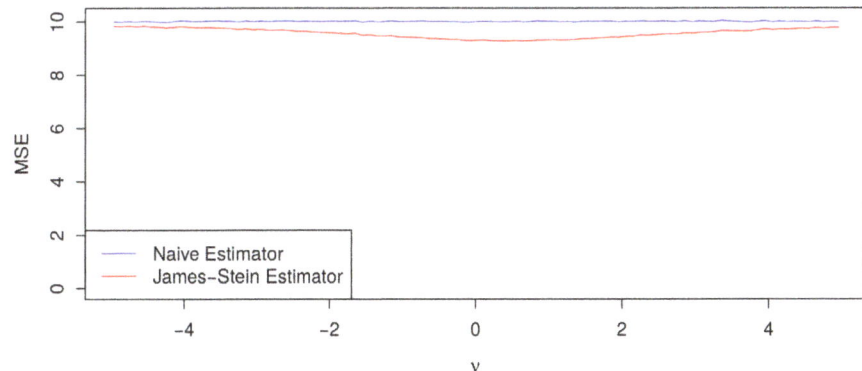

Fig. 8. MSE for the naive estimator (blue) and James-Stein Estimator (red) when the θ_j are evenly distributed on [-4,5]. The James-Stein Estimator dominates the naive estimator regardless of the choice of shrinkage term ν, though the difference is less pronounced than in Figure 7. This is because here the variance of the θ_j is larger (due to the θ_j being distributed over a larger range), which results in less shrinkage.

Note the similarity between (3.2) and (4.2): $E(\theta_j \mid Y, \phi)$ is also a shrinkage estimator of θ_j. This is an important result as it implies that Bayesian hierarchical models induce shrinking, as with the James-Stein Estimators, but also automatically pick the best ν (i.e. μ in (4.1)) and provide model-based estimates of ϕ. The Bayesian perspective, however, requires that prior distributions be specified for σ^2, τ^2, and μ.

As an extended example, suppose we wish to infer color correction parameters for a sample of 288 Type Ia supernova (SNIa) light curves from Kessler et al. (2009). SNIa light curves vary systematically across color bands, and the color correction parameter, denoted c_j for supernova j, describes how flux depends on color. The color correction that is observed, with error, is denoted $\hat{c}_j$. A hierarchical model for inferring the color correction parameters for a collection of supernovae (SNe) is

$$\hat{c}_j | c_j \overset{\text{indep}}{\sim} \text{NORM}(c_j, \sigma_j^2) \quad \text{for} \quad j = 1, \ldots, 288 \tag{4.3}$$

with

$$c_j \overset{\text{indep}}{\sim} \text{NORM}(c_0, R_c^2) \quad \text{and} \quad p(c_0, R_c) \propto 1, \tag{4.4}$$

where the object-specific measurement variances, σ_j^2, are assumed known and R_c^2 is the population variance of the colors c_j.

A naive object-wise likelihood-based estimate for each c_j is $\hat{c}_j \pm \sigma_j$. We aim to compare these naive estimates with the shrinkage estimates arising from the hierarchical model fit with a Gibbs sampler. To derive the Gibbs sampler we note:

(a) Given (c_0, R_C^2), the c_j are mutually independent and we can derive their marginal posterior distribution. Specifically,

$$c_j \mid (\hat{c}_j, c_0, R_C^2) \overset{\text{indep}}{\sim} \text{NORM}\left(\mu_j,\ s_j^2\right),$$

where $\mu_j = \left(\frac{\hat{c}_j}{\sigma_j^2} + \frac{c_0}{R_C^2}\right) \Big/ \left(\frac{1}{\sigma_j^2} + \frac{1}{R_C^2}\right)$ and $s_j^2 = \left(\frac{1}{\sigma^2} + \frac{1}{R_C^2}\right)^{-1}$.

(b) Given $c_1, \ldots, c_G$, we can derive the joint posterior distributions of (c_0, R_C^2). Specifically,

$$R_C^2 \mid (c_1, \ldots, c_G) \sim \frac{\sum_{j=1}^{G}(c_j - \bar{c})^2}{\chi_{G-2}^2} \quad \text{with} \quad \bar{c} = \frac{1}{G}\sum_{j=1}^{G} c_j,\ \text{and}$$

$$c_0 \mid (c_1, \ldots, c_G), R_C^2 \sim \text{NORM}\left(\bar{c},\ R_C^2/G\right).$$

We can exploit (a) and (b) to construct a two-step Gibbs sampler. It proceeds as follows after drawing $c_0^{(0)}$, $(R_C^2)^{(0)}$, and $(c_1^{(0)}, \ldots, c_G^{(0)})$ from some starting distribution. For $t = 1, 2, 3, \ldots$

Step 1. Sample $c_1, \ldots c_G$ from their joint posterior given (c_0, R_C^2):

$$c_j^{(t)} \mid (\hat{c}_j, c_0^{(t-1)}, (R_C^2)^{(t-1)}) \overset{\text{indep}}{\sim} \text{NORM}\left(\mu_j,\ s_j^2\right),$$

where $\mu_j = \left(\frac{\hat{c}_j}{\sigma_j^2} + \frac{c_0^{(t-1)}}{(R_C^2)^{(t-1)}}\right) \Big/ \left(\frac{1}{\sigma_j^2} + \frac{1}{(R_C^2)^{(t-1)}}\right)$ and $s_j^2 = \left(\frac{1}{\sigma^2} + \frac{1}{(R_C^2)^{(t-1)}}\right)$

Step 2. Sample (c_0, R_C^2) from their joint posterior given $c_1, \ldots c_G$:

$$(R_C^2)^{(t)} \mid (c_1^{(t)}, \ldots, c_G^{(t)}) \sim \frac{\sum_{j=1}^{G}(c_j^{(t)} - \bar{c})^2}{\chi_{G-2}^2} \quad \text{with} \quad \bar{c} = \frac{1}{G}\sum_{j=1}^{G} c_j^{(t)},\ \text{and}$$

$$c_0^{(t)} \mid (c_1^{(t)}, \ldots, c_G^{(t)}), (R_C^2)^{(t)} \sim \text{NORM}\left(\bar{c},\ (R_C^2)^{(t)}/G\right).$$

The hierarchical model in (4.3)-(4.4) fitted with this Gibbs sampler is summarized in Figures 9 and 10. In Figure 9 the conditional posterior expectations of the color correction parameters for a selection of Type Ia SNe are plotted as a function of R_c, the prior standard deviation on the c_j (i.e., the population standard deviation on the c_j). The different colors represent the different Type Ia SNe, with the plusses showing the fitted values of c_j obtained from an object-wise likelihood fit (i.e., $\hat{c}_j$). Notice that as the prior standard deviation R_c decreases, the conditional posterior expectation of each c_j shrinks toward the center. The vertical dashed lines with grey shading in between portray a 95% Bayesian credible interval for R_c,

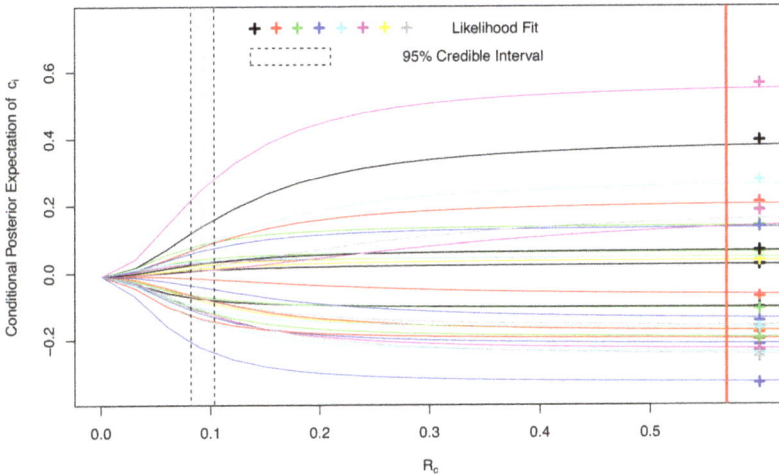

Fig. 9. Shrinkage of the fitted color correction in the simple hierarchical model in (4.3)-(4.4). Different colors represent different SNe. The "+" symbols indicate the fitted values of each c_j via an object-wise likelihood fit. As the prior standard deviation R_c decreases, the conditional posterior expectations of the c_j shrink toward the center.

and so the fitted values for the color correction parameters are (roughly) the values the colored curves take in the middle of the grey shading. Thus using a hierarchical model, which pools information across the different SNe, dramatically changes the fits. A similar effect is seen for the conditional posterior standard deviations of each c_j, as illustrated in Figure 10. When such pooling occurs we say that the model "borrows strength" to obtain more precise estimates. More specifically, while these estimates may be more biased than their non-pooled counterparts, there is enough reduction in variance for the pooled estimates to have lower MSE via the bias-variance tradeoff. Recall that MSE = bias^2 + variance. In this way, Bayesian hierarchical models may produce *biased* estimates that nevertheless have better *frequentist* properties than the corresponding naive non-Bayesian estimators.

5   Concluding Remarks

A particular advantage of the Bayesian perspective is that the gain from James-Stein Estimation is automatic; James and Stein had to find their estimator and derive its properties! The Bayesian estimator follows automatically by applying Bayes' Theorem. Bayesians have a method to generate estimators, and the general principle is easily tailored to any specific scientific problem. A caution is that results can depend on prior distributions for parameters that reside deep within the model, and far from the data. It is therefore recommended to check the

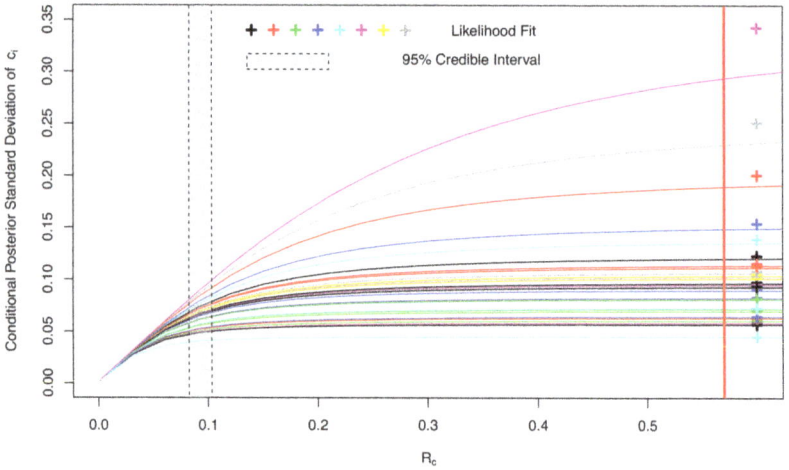

Fig. 10. Standard deviation of the fitted color correction. The "+" symbols indicate the fitted values under an object-wise likelihood fit. The hierarchical model in (4.3)-(4.4) "borrows strength" across data sets for more precise estimates. That is, the estimates' conditional posterior standard deviations under the hierarchical model, which fall within the 95% credible interval above, are less than or equal to those under the object-wise likelihood fit.

sensitivity of the results to any reasonable choice of prior distribution.

The models discussed in this chapter are tailored to specific examples at hand, but a search of the literature reveals a multitude of Bayesian hierarchical models being used in a variety of astrophysics applications. These include, but are not limited to: examining the solar cycle (Yu et al., 2012), modeling exoplanet populations (Wolfgang et al., 2016), characterizing multiple populations in globular clusters (Stenning et al., 2016; Wagner-Kaiser et al., 2016a,b), deriving galaxy redshift distributions from photometric surveys (Leistedt et al., 2016), and inferring cosmological parameters (Alsing et al., 2017). These examples are certainly not exhaustive; we merely attempt to illustrate the use of Bayesian hierarchical models in many areas of astronomy and astrophysics. Combined with the computational techniques discussed in Chapter 2, the general methodologies described in this chapter should allow the reader to construct and fit complex Bayesian hierarchical models for their particular astrophysics problems. Those desiring a more in-depth treatment of Bayesian hierarchical models are encouraged to consult, for example, Gelman et al. (2013) and references therein. In addition, Chapter 5 of this volume addresses clustering of globulars via Bayesian nonparametric hierarchical models.

Acknowledgements

The authors would like to thank Dr. Xiyun Jiao for implementing the numerical studies described in Figures 1, 2, 3, 4, 9 and 10, and for producing the figures themselves.

References

Alsing, J., Heavens, A., and Jaffe, A. (2017). Cosmological Parameters, Shear Maps and Power Spectra from CFHTLenS Using Bayesian Hierarchical Inference. *Monthly Notices of the Royal Astronomical Society*, 466, (3): 3272–3292.

Gelman, A., Carlin, J. B., Stern, H. S., Dunson, D. B., Vehtari, A., and Rubin, D. B. *Bayesian Data Analysis: Third Edition* (Chapman & Hall/CRC Press, Boca Raton, London, New York, 2013)

Kessler, R., et al. (2009). First-Year Sloan Digital Sky Survey-II Supernova Results: Hubble Diagram and Cosmological Parameters. *The Astrophysical Journal Supplement*, 185, (1): 32–84.

Leistedt, B., Mortlock, D. J., and Peiris, H.V. (2016). Hierarchical Bayesian Inference of Galaxy Redshift Distributions from Photometric Surveys. *Monthly Notices of the Royal Astronomical Society*, 460, (4): 4258–4267.

Stenning, D. C., Wagner-Kaiser, R., Robinson, E., van Dyk, D.A., von Hippel, T., Sarajedini, A., and Stein, N. (2016). Bayesian Analysis of Two Stellar Populations in Galactic Globular Clusters. I. Statistical and Computational Methods. *The Astrophysical Journal*, 826, (1): 16pp.

Wagner-Kaiser, R., Stenning, D.C., Robinson, E., von Hippel, T., van Dyk, D.A., Sarajedini, A., Stein, N., and Jeffreys, W.H. (2016). Bayesian Analysis of Two Stellar Populations in Galactic Globular Clusters. II. NGC 5024, NGC 5272, and NGC 6352. *The Astrophysical Journal*, 826, (1): 18pp.

Wagner-Kaiser, R., Stenning, D.C., Sarajedini, A., von Hippel, T., von Hippel, T., van Dyk, D.A., Robinson, EA., Stein, N., and Jeffreys, W.H. (2016). Bayesian Analysis of Two Stellar Populations in Galactic Globular Clusters. III. Analysis of 30 clusters. *Monthy Notices of the Royal Astronomical Society*, 463, (4): 3768–3782.

Wolfgang, A., Rogers, L.A., and Ford, E.B. (2016). Probabilistic Mass-Radius Relationship for Sub-Neptune-Sized Planets. *The Astrophysical Journal*, 825, (1): 14pp.

Yu, Y., van Dyk, D.A., Kashyap, V., and Young, C. A. (2012). A Bayesian Analysis of the Correlations Among Sunspot Cycles. *Solar Physics*, 281, (2): 847–862.

APPROXIMATE BAYESIAN COMPUTATION, AN INTRODUCTION

Christian P. Robert [1]

Abstract. Approximate Bayesian Computation (ABC) methods have become a "mainstream" statistical technique in the past decade, following the realisation that they were a form of non-parametric inference, and connected as well with the econometric technique of indirect inference. In this survey of ABC methods, we focus on the basics of ABC and cover some of the recent literature, following our earlier survey in Marin et al. (2011). Given the recent paradigm shift in the perception and practice of ABC model choice, we insist on this aspect of ABC techniques, including in addition some convergence results.

1 Mudmap: ABC at a glance

While statistical (probabilistic) models are always to be held at a critical distance, being at best approximations of real phenomena (Box, 1959, Gelman et al., 2013), and while more complex statistical models are not necessarily better representations of those phenomena, it is becoming increasingly the case that the complexity of models is a barrier to the most common tools in statistical analysis. Complexity might stem from many possible causes, from an elaborate description of the randomness behind the phenomenon to be analysed, to the handling of massive amounts of observations, to imperatives for real-time analysis, to considerable percentages of missing data in an otherwise regular model. All those reasons contribute to make computing the likelihood function a formidable task.

Example 1. *Kingman's coalescent* Inference in population genetic relies on models such as Kingman's coalescent trees. This is a representative example of cases when the likelihood function associated with the data cannot be computed in a manageable time (Tavaré et al., 1997; Beaumont et al., 2002; Cornuet et al., 2008). The fundamental reason for this impossibility is that the statistical model associated with coalescent data needs to integrate over a space of trees of high complexity.

Kingman's coalescent trees are probabilistic models representing the evolution from an unobserved common ancestor to a collection of N (observed) populations, described by the

---

[1] CEREMADE, Université Paris Dauphine PSL, & Department of Statistics, University of Warwick
e-mail: xian@ceremade.dauphine.fr

frequencies of genetic traits such as alleles of loci from the genome in single nucleotide polymorphism (SNP) datasets. For two populations j_1 and j_2 and a given locus, at current time, with allele sizes x_{j_1} and x_{j_2}, a binary tree has for root the most recent time in the past for which they have a common ancestor, defined as the coalescence time τ_{j_1,j_2}. The two copies are thus separated by a branch of gene genealogy of total length $2\tau_{j_1,j_2}$. As explained in Slatkin (1995), according to Kingman's coalescent process, during that duration $2\tau_{j_1,j_2}$, the number of mutations is a random variable distributed from a Poisson distribution with parameter $2\mu\tau_{j_1,j_2}$. Aggregating all populations by pairing the most recently diverged pairs and repeating the pairing on the most recent common ancestors of those pairs produce a binary tree which root is the most recent common ancestor of the collection of populations and with as many branches as there are populations. For a given tree topology, such as the one provided in Figure 1, inferring the tree parameters (coalescent times and mutation rate) is a challenge, because the likelihood of the observations is not available, while the likelihood of the completed model involves missing variables, namely the $2(N-1)$ mutations along all edges of the graph. While this is not a large dimension issue in the case of Figure 1, building an efficient completion mecchanism such as data augmentation or importance sampling proves to be quite complicated (Stephens and Donnelly, 2000). ◄

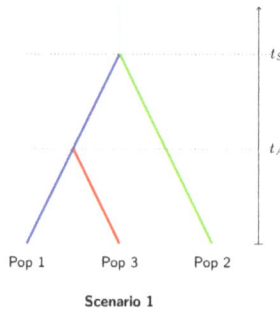

Fig. 1. Possible model of historical relationships between three populations of a given species *(Source: Pudlo et al. (2015), with permission).*

Example 2. *Lotka–Volterra prey-predator model* The Lotka-Volterra model (Wilkinson, 2006) describes interactions between a first species, referred to as the prey species, and a second species, referred to as the predator species, in terms of population sizes, $x_1(t)$ and $x_2(t)$. Given the parameter $\theta = (\theta_1, \theta_2)$, the model on the respective population sizes is driven by a system of differential equations (ODEs):

$$\frac{dx_1}{dt} = \theta_1 x_1 - x_1 x_2,$$

$$\frac{dx_2}{dt} = \theta_2 x_1 x_2 - x_2.$$

with initial values $(x_1(0), x_2(0))$. Typically, one does not observe the entire curve $\mathbf{x}(t)$, but only at a finite number of times $t_1, \ldots, t_R$. Furthermore, the $\mathbf{x}(t_i)$'s are measured with error,

$$\mathbf{y}(t_i) = \mathbf{x}(t_i) + \mathbf{v}(t_i),$$

where $v(t_i) \overset{\text{i.i.d.}}{\sim} \mathcal{N}_2(0, \Sigma_v)$, a bivariate Normal distribution with covariance matrix Σ_v.

While the likelihood associated with this collection of observations is intractable, the process $\mathbf{y}$ can be simulated, which means particle MCMC solutions exist, as in Golightly and Wilkinson (2011), even though an ABC implementation is more straightforward (Toni et al., 2009). ◀

In such situations, statisticians will try to provide answers by

- modifying the original model towards a more manageable version (see, e.g., the variational Bayes literature as in Jaakkola and Jordan, 2000);

- using only relevant aspects of the model (see, e.g., the many versions of the method of moments in econometrics, Gouriéroux *et al.*, 1993, Heggland and Frigessi, 2004, Gallant and Tauchen, 1996);

- constructing new tools (with numerous examples ranging from the EM algorithm, Dempster et al., 1977, to machine learning, Breiman et al., 1984; Hastie et al., 2001).

The Approximate Bayesian computation (ABC) method covered by this chapter is a mix of these different solutions in that it does not return an answer to the original question (namely, to deliver the posterior distribution associated with the original dataset and the postulated model(s)), only uses some summaries of the data (even though it most often requires a constructive definition of the model), and construct a simulation tool that is to some extent novel (albeit a rudimentary version of accept-reject algorithm, Robert and Casella, 2004).

Without yet attempting to justify the ABC method from a theoretical perspective, let me provide here a quick and handwaving description of its *modus vivendi* (operational mode). Its goal is to substitute a Monte Carlo approach to an intractable posterior distribution, while preserving the Bayesian aspect of the statistical analysis. The ABC method is based on the intuition that if a parameter value θ produces one or several simulated datasets $x(\theta)$ that are resembling[1] the observed data set $x(\theta^0)$ then θ must be close to the value of the parameter that stands behind the data, θ^0. And, conversely, if the simulated dataset $x(\theta)$ differs a lot from the observed data $x(\theta^0)$, then θ is presumably different from θ^0. ABC goes beyond this intuition by quantifying the resemblance and estimating the bound for the datasets to be "close enough". Starting from a collection of parameter values usually generated from the prior, ABC first associates with each value a simulated dataset of

---

[1] The notation $x(\theta)$ is stressing the point that the simulated data is a random transform of the parameter θ, $x(\theta) \sim f(\cdot|\theta)$. From a programming perspective, $x(\theta)$ actually is a function of θ and of a random variable or sequence, o: $x(\theta, o)$. By extension, assuming the postulated model is correct, there exists a "true" value of the parameter, θ^o, such that the observed data writes as $x(\theta^0)$.

the same nature as the observed one, i.e., a pseudo-dataset, and then rejects values of θ for which the simulated dataset is too unlike the observed one. The surviving parameter values are subsequently used as if they were generated from the posterior distribution, even though they are not, due to several reasons discussed below. The major reason for failing to accommodate for this difference is that the approximation effect is difficult (and costly) to evaluate, even though the convergence rates of ABC are now well-characterised (Li and Fearnhead, 2018; Frazier et al., 2018).

While ABC is rarely fast, due to the reason that many simulated samples need to be produced and that the underlying statistical model is complex enough to lead to costly generations, it often is the unique answer to settings where regular Monte Carlo methods (including MCMC, Robert and Casella, 2004, and particle filters, Doucet et al., 1999) fail. The method is easily parallelisable as well as applicable to sequential settings, due to its rudimentary nature. Furthermore, once the (massive) collection of pairs $(\theta, x(\theta))$ is produced, it can be exploited multiple times, which makes it paradoxically available for some real time applications.

2   ABC Basics

2.1   Intractable likelihoods

Although it has now spread to a wide range of application domains, Approximate Bayesian Computation (ABC) was first introduced in population genetics (Tavaré et al., 1997; Pritchard et al., 1999) to handle models with intractable likelihoods (Beaumont, 2010). By *intractable*, we mean models where the likelihood function $\ell(\theta|y)$

- is completely defined by the probabilistic model, $y \sim f(y|\theta)$;

- is not available in closed form, neither is it numerical derivation;

- cannot easily be either completed or demarginalised (see, e.g. Tanner and Wong, 1987; Robert and Casella, 2004);

- cannot be estimated by an unbiased estimator (Andrieu and Roberts, 2009).

This intractability prohibits the direct implementation of a generic MCMC algorithm like Gibbs or Metropolis–Hastings schemes. Examples of latent variable models of high dimension abound, primarily in population genetics, but more generally in models including combinatorial structures (e.g., trees, graphs), intractable normalising constants as in $f(y|\theta) = g(y|\theta)/Z(\theta)$ (e.g. Markov random fields, exponential graphs) and other missing (or latent) variables, i.e. when

$$f(y|\theta) = \int_{\mathscr{G}} f(y|G, \theta) f(G|\theta) \mathrm{d}G$$

cannot be computed in a manageable way (while $f(y|G, \theta)$ and $f(G|\theta)$ are available).

Example 3. As an intuitive try to intractable likelihoods, consider the "case of the single socks", as proposed by Rasmus Bååth on his blog. At the end of one's laundry, the 11 first socks extracted from the washing machine are single (meaning that no complete pair is recovered). *What is the posterior distribution on the number of socks? and on the number of pairs?*

This sounds like an impossible task, but it can be solved by setting a prior on the number of socks, n_s, chosen to be a negative binomial $\mathscr{N}eg(N,\rho)$ random variable based on the size of the family, with mean 30 and standard deviation 15, and on the proportion of pairs in the laundry, chosen to derived from a Beta $p \sim \mathscr{B}e(15,2)$ weight to reflect on the low proportion of single socks, namely $n_p = \lceil p n_s/2 \rceil$, the integral part by excess of $p n_s/2$. Given (n_s, n_p), it is then straightforward to generate a laundry sequence of 11 socks by a simple sampling without replacement from the population of socks. *A contrario*, it is much more involved to express the distribution of the number of single socks in those 11 random draws (albeit possible, see below). ◄

The idea of the approximation behind ABC is both surprisingly simple and fundamentally related to the very nature of statistics, i.e., as solving an inverse problem. Indeed, ABC relies on the feasibility of producing simulated (parameters and) data from the inferred model or models, as it evaluates the unavailable likelihood by the proximity of this simulated data to the observed data. In other words, it relies on the natural assumption that the *forward* step induced by the probabilistic model—from model to data—is reasonably easy to implement in contrast with the *backward* step—from data to model.

2.2 An exact Bayesian computation

> "Bayesian statistics and Monte Carlo methods are ideally suited to the task of passing many models over one dataset." D. Rubin, 1984

Not so coincidentally, Rubin (1984), quoted above, used this representation as a mostly non-algorithmic motivation for conducting Bayesian analysis (as opposed to other forms of inference). This paper indeed details the accept-reject algorithm (Robert and Casella, 2004) at the core of the ABC algorithm. Namely, the following algorithm

---

Algorithm 1 Accept-reject for Bayesian analysis

---
repeat
  Generate $\theta \sim \pi(\theta)$;
  Generate $x \sim f(x|\theta)$;
  Accept θ if $x = x_0$
until acceptance
return the accepted value of θ

---

returns as accepted value an output *exactly* generated from the posterior distribution, $\pi(\theta|x_0)$.

Example 4. If we return to the socks example 3, running Algorithm 1 means running the following steps
repeat
  Generate $n_s \sim \mathscr{N}eg(N,\rho)$ and $p \sim \mathscr{B}e(15,2)$

Set $n_p = \lceil p n_s / 2 \rceil$
Sample 11 terms from $\{o_{11}, o_{12}, \ldots, o_{n_p 1}, o_{n_p 2}, s_1, \ldots, s_{n_s - 2n_p}\}$
Accept (n_s, n_p) if there is no pair (o_{i1}, o_{i2}) in the sample
until acceptance
return the accepted value of (n_s, n_p)

and this loop will produce an output from the posterior distribution of (n_s, n_p), that is, conditional on the event that no pair occurred out of 11 draws without replacement. Running the implementation of the above algorithm as done in Bååth's R code leads to Fig. 2. The number of proposals in the above loop was 10^5, resulting in above 10^4 acceptances.

As mentionned above, the probability that the 11 socks all come from different pairs can be computed by introducing a latent variable, namely the number k of orphan socks in the 11 socks, out of the $n_s - 2n_p$ existing orphans. Integrating out this latent variable k (and using Feller, 1970, Chap. 2, Exercise 26 for the number of different pairs in the remaining $11 - k$ socks), leads to

$$\sum_{k=0}^{11} \frac{\binom{n_s - 2n_p}{k} \binom{2n_p}{11-k}}{\binom{n_s}{11}} \cdot \frac{2^{11-k} \binom{n_p}{11-k}}{\binom{2n_p}{11-k}}$$

as the probability of returning no complete pair out of the 11 draws. If we discretise the Beta $\mathscr{B}e(15, 2)$ distribution to obtain the probability mass function of n_p, we are therefore endowed with a closed-form posterior distribution

$$\pi(n_s, n_p | \mathscr{D}) \propto \binom{n_s + N - 1}{n_s} \rho^{n_s} \int_{2n_p/n_s}^{2(n_p+1)/n_s} p^{16}(1-p)^3 \, dp \sum_{k=0}^{11} \frac{\binom{n_s - 2n_p}{k} \binom{2n_p}{11-k}}{\binom{n_s}{11}} \cdot \frac{2^{11-k} \binom{n_p}{11-k}}{\binom{2n_p}{11-k}}$$

that we can compare with the ABC output (even though this ABC output is guaranteed to correspond to simulations from the true posterior distribution[2]). As demonstrated in Fig. 3, there is indeed a perfect fit between the simulations and the target.

Obviously, a slight extension of this setup with, say, a second type of socks or a different rule for pulling the socks out of the washing machine (or out of different laundry bags) could sufficiently complicate the likelihood associated with the observations to reach intractability. See for instance Arratia and DeSalvo (2012) for an interesting alternative of Feller's (1970) shoes cupboard problem.   ◀

2.3   Enters the approximation

Now, ABC proceeds one step further in the approximation, replacing the acceptance step with the tolerance condition

$$d(x, x_0) < \varepsilon$$

in order to handle continuous (and large finite) sampling spaces, $\mathfrak{X}$, but this early occurrence in Rubin (1984) is definitely worth signalling. It is also relevant that Rubin does not

---

[2]This feature means that the 'A' in 'ABC' is superfluous in this special case!

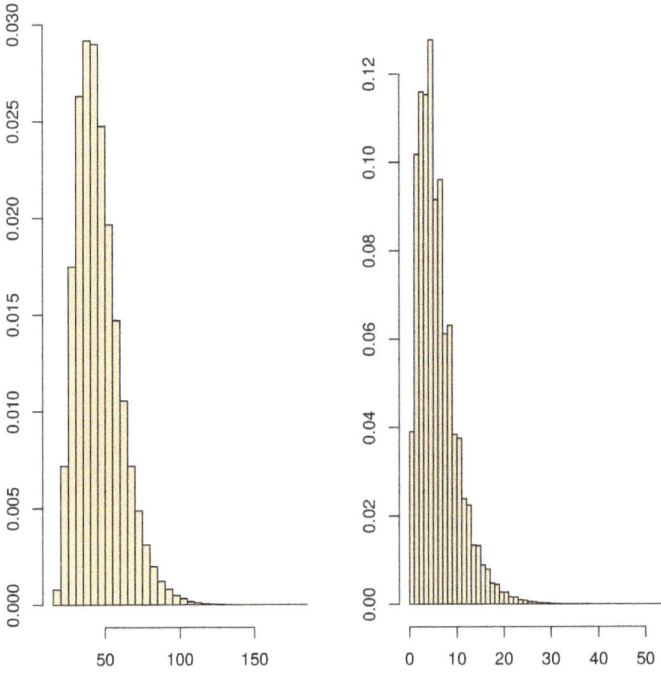

Fig. 2. ABC based simulation of posterior distributions on *(left)* the number of socks, n_s, and *(right)* the number of odd socks, n_o, relying on 10^6 proposals (n_s, n_p) simulated from the prior. *(R code kindly provided by Rasmus Bååth).*

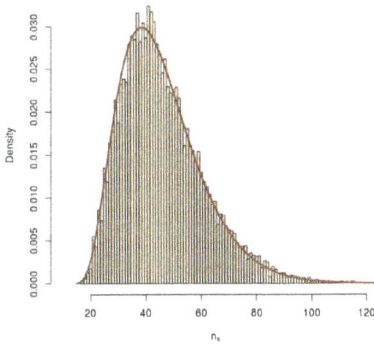

Fig. 3. Same graphs as in Fig. 2 with true posterior marginal distributions in superposition.

promote this simulation method in situations where the likelihood is not available but rather as an intuitive way to understand posterior distributions from a frequentist perspective, because θ's from the posterior are those that could have generated the observed data. (The issue of the zero probability of the exact equality between simulated and observed data is not dealt with in the paper, maybe because the notion of a "match" between simulated and observed data is not clearly defined.) Another (just as) early occurrence of an ABC-like algorithm was proposed by Diggle and Gratton (1984).

Algorithm 2 ABC (basic version)

---

   for $t = 1$ to N **do**
      repeat
         Generate θ^* from the prior $\pi(\cdot)$
         Generate x^* from the model $f(\cdot|\theta^*)$
         Compute the distance $\rho(\mathbf{x}^0, \mathbf{x}^*)$
         Accept θ^* if $\rho(\mathbf{x}^0, \mathbf{x}^*) < \varepsilon$
      until acceptance
   end for
   return N accepted values of θ^*

    The ABC method is formally implemented as in Algorithm 2, which requires calibrating the objects $\rho(\cdot, \cdot)$, called the *distance* or *divergence* measure, N, number of accepted simulations, and ε, called the *tolerance*. Algorithm 2 is exact (in the sense of Algorithm 1) when $\varepsilon = 0$. However, this is at best a formal remark since this ideal setting cannot be found in most problems where ABC is needed (see Grelaud et al. (2009) for a specific counterexample) and a positive tolerance is required in practical settings[3]. While several approaches are found in the literature, we follow here the practice of selecting ε as a quantile of the simulated distances $\rho(\mathbf{x}^0, \mathbf{x}^*)$, which turns out to express ABC as a k-nn method, as pointed out by Biau et al. (2015) and discussed in Section 3.1.

    This algorithm is easy to call when checking the performances of the ABC methods on toy examples where the exact posterior distribution is known, in order to test the impact of the various calibration parameters. See for instance Marin et al. (2011) with the case of the MA(2) model. We illustrate the behaviour of the algorithm in a slightly more challenging setting.

Example 5. A surprisingly complex probability density (and hence likelihood function) is the one associated with the empirical mean $\bar{x}_n$ of a Student's t sample.[4] Indeed, if

$$(x_1, \ldots, x_n) \overset{\text{i.i.d.}}{\sim} \mathfrak{T}(\nu, \mu, \tau),$$

---

[3]There even are arguments, see e.g. Fearnhead and Prangle (2012), to justify positive values of ε as preferable.
[4]We are aware that there exist two differing definitions for the $\mathfrak{T}(\nu, \mu, \tau)$ distribution. One is considering μ and τ as location and scale parameters: this is the interpretation chosen in this example. Another one starts from a $\mathcal{N}(\mu, \tau)$ variate and divides it by a normalised χ_ν variate, which leads to a non-standard density for even a single variable.

the resulting $\bar{x}_n$ has no standard distribution, even though it is also a location scale distribution with parameters μ and τ. To see this, consider that $x_i = \mu + \tau\xi_i$, with $\xi_i \sim \mathfrak{T}_v$. Then

$$\bar{x}_n = \mu + \tau\bar{\xi}_n, \tag{2.1}$$

with $\bar{\xi}_n$ distributed from a density that cannot be expressed otherwise than as an $(n-1)$-convolution of t's.

If we observe $p \geq 1$ realisations of $\bar{x}_n$, denoted $\bar{x}^1,\ldots,\bar{x}^p$, Algorithm 2 may be the solution to handling the corresponding implicit likelihood. When the prior on (μ, τ) is the normal-gamma prior

$$\tau^{-1} \sim \mathscr{G}a(1,1), \; \mu|\tau \sim \mathscr{N}(0,2\tau),$$

Algorithm 2 consists in

1. generating a large sample of (μ, τ) from this prior (the sample is often called a *reference table*, then

2. generating a pseudo-sample $(\bar{x}^1,\ldots,\bar{x}^p)$ for each pair (μ, τ) in the reference table, and

3. deriving the distances ρ between pseudo- and true samples.

The reference table is then post-processed by keeping the parameter values that lead to the $(100\varepsilon)\%$ smallest distances. The choice of the distance is arbitrary, it could for instance be the average squared error

$$\rho(\mathbf{x}_0, \mathbf{x}_1) = \sum_{i=1}^{p} (\bar{x}_0^i - \bar{x}_1^i)^2.$$

Figure 4 compares the parameter values selected by Algorithm 2 with the entire reference table, based on 10^5 simulations. The concentration of the parameter values near the true value $(0,1)$ is noticeable, albeit with a skewness towards the smallest values of τ that may reflect the strong skewness in the prior distribution on τ.   ◄

A further difficulty arises when the prior on θ is improper and hence cannot be simulated. It is then impossible to use directly Algorithm 2. Instead, we can proceed by either

1. using a proper prior with a very large variance, à la BUGS (Lunn et al., 2010), but this is a very inefficient and wasteful proposal, both *per se* and in this particular setting, since most values generated from this prior will be fully incompatible with the data; or

2. replacing the prior simulation by a simulation from a pseudo-posterior, based on the data and mimicking to some extent the true posterior distribution, and weighting the outcome by the ratio of the prior over the pseudo-posterior. In the case of Example 5, we could instead use the posterior distribution associated with a normal sample, that is, by pretending the sample of the $\bar{x}_n$'s is made of normal observations with mean μ and variances $(v/v-2)(\tau^2/n)$; or

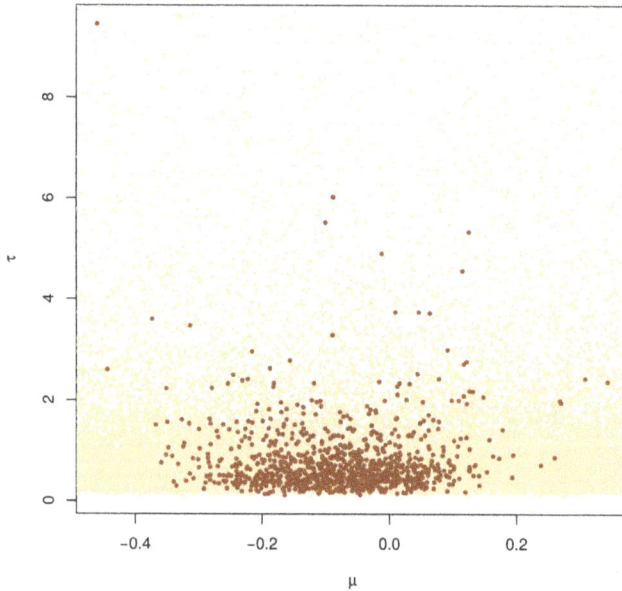

Fig. 4. Sample of $1,000$ simulations from Algorithm 2 when the data is made of 10 t averages with the same sample size $n = 21$ and when the 10^5 ABC simulations in the background constitute the reference table, taken from the normal-gamma prior. *(Note: the true value of (μ, τ) is $(0,1)$.)*

3. using part of the data to build a simpler but achievable posterior distribution. This solution is not available for Example 5, since even a single $\bar{x}^i$ is associated with a complex density; or

4. introducing latent variables to recover a closed form conditional posterior. In the setting of Example 5, it would prove very dear, since this requires producing n pairs $(y_j, z_j) \sim \mathcal{N}(0,1) \times \chi_\nu^2$ to decompose $\bar{x}_{n_i}$ as

$$\bar{x}_n = \mu + \tau^{1/2} \frac{1}{n} \sum_{j=1}^{n} y_j / \sqrt{z_j/\nu}.$$

Example 6. Let us thus consider the more (when compared with Example 5) challenging case where (a) we observed independently $\bar{x}_{n_1}, \ldots, \bar{x}_{n_p}$ that all are averages of Student's t samples with different sample sizes $n_1, \ldots, n_p$, and (b) the prior on (μ, τ) is the reference prior $\pi(\mu, \tau) = 1/\tau$.

We select the second solution proposed above, namely to rely on a normal approximation for the distribution of the observations, in order to build the following proposal:

$$(\mu, \tau^{-2})|\bar{x}_{n_1}, \ldots, \bar{x}_{n_p} \sim \mathcal{N}\left(\bar{\bar{x}}, v\tau^2/(v-2)\Sigma n_i\right) \times \mathcal{G}\left(1 + p/2, (v-2)s^2/2v\right)$$

that serves as a proxy generator in the first step of the above algorithm.

If we apply Algorithm 2 to this problem, due to the representation (2.1), we can follow the next steps:

> **for** $t = 1$ to N **do**
>> Generate θ^* from the pseudo-posterior $\pi^*(\cdot|x)$
>> Create a sample $(\bar{\xi}_{n_1}^t, \ldots, \bar{\xi}_{n_p}^t)$
>> Derive the transform $x^t = (\bar{x}_{n_1}^t, \ldots, \bar{x}_{n_p}^t)$
>> Compute the distance $\rho(x^0, x^t)$
>> Accept θ^* if $\rho(x^0, x^t) < \varepsilon$
>
> **end for**
> **return** N accepted values of θ^* along with importance weights $\omega^* \propto \pi(\theta^*)/\pi^*(\theta^*|x)$
> or resample those θ^* with replacement according to the corresponding ω^*'s

where the distance is again arbitrarily chosen as the sum of the weighted squared differences.

Following this algorithm for a dataset of 10 averages simulated from central t distributions (i.e., with $\mu = 0$, $\tau = 1$), we obtain an ABC sample displayed on Fig. 5, which shows a reasonable variability of the sample around the true value $(0,1)$. The 10^3 points indicated on this picture are the output of a weighted resampling. If we compare the θ^*'s simulated from the pseudo-posterior $\pi^*(\cdot|x)$ with those finally sampled, the difference is quite limited, as exhibited in Fig. 6. The selected points do remain in a close neighbourhood of the mode. This behaviour remains constant through the choice of ε, so we can attribute it to (at least) two possible reasons. The first explanation is that the likelihood associated with $\bar{x}_{n_i}$ should be quite close to a normal likelihood, hence that the pseudo-posterior provides a fairly accurate representation of the true posterior distribution. The second explanation is a *contrario* that the ABC output reflects a lack of discrimination in the condition $\rho(x^0, x^t) < \varepsilon$, even for small values of ε and hence corresponds to simulations from a pseudo-posterior that differs from the posterior.   ◄

2.4   Enter the summaries

In realistic settings, Algorithm 2 is almost never ever used as such, due to the curse of dimensionality. Indeed, the data x^0 is generally complex enough for the proximity measure $\rho(x^0, x^*)$ to be far from small (and hence discriminating). As illustrated on the time series (toy) example of Marin et al. (2011), the signal-to-noise ratio produced by $\rho(x^0, x^*) < \varepsilon$ falls dramatically as the dimension (of the data) increases. This means a corresponding increase in either the total number of simulations N_{ref} or in the tolerance ε is required to preserve a positive acceptance rate. In other words, we are aiming at the parameters to be close rather than at the observations themselves. In practice, it is thus paramount to first summarise the data (and decrease the dimension) in a so-called *summary statistic* before computing a proximity index. Thus enters the notion of *summary statistic*[5] that is central

---

[5]While, for a statistician, a statistic is by nature a *summary* of the data, hence making the term redundant, the non-statisticians who introduced this notion in the ABC algorithms felt the need to stress this aspect.

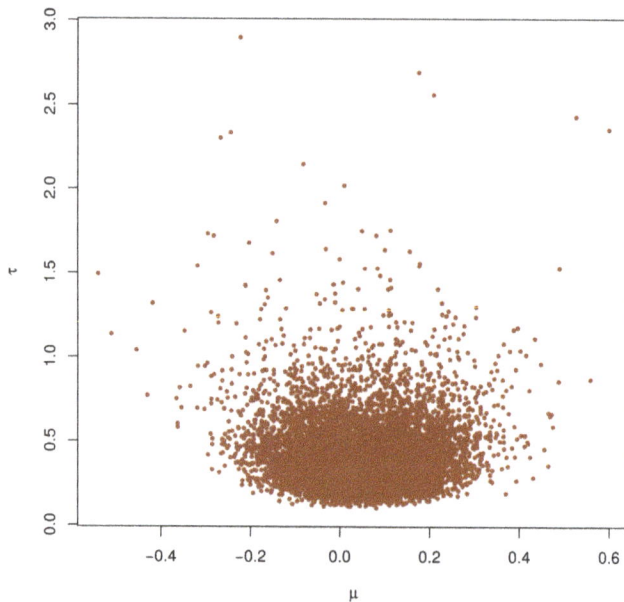

Fig. 5. Sample of $1,000$ simulations from Algorithm 2 when the data is made of 10 t averages with respective sample sizes $n_i = 9,8,8,11,10,5,4,3,5,3$, and when the 10^6 ABC simulations are taken from a pseudo-posterior. *(Note: the true value of (μ, τ) is $(0,1)$.)*

to operational ABC algorithms, as well as the subject of much debate, as discussed in Marin et al. (2011); Blum et al. (2013) and below. A more realistic version of the ABC algorithm is produced in Algorithm 3, where $S(\cdot)$ denotes the summary statistic.

Algorithm 3 ABC (version with summary)

---

for $t = 1$ to N_{ref} **do**

    Generate $\theta^{(t)}$ from the prior $\pi(\cdot)$

    Generate $\boldsymbol{x}^{(t)}$ from the model $f(\cdot|\theta^{(t)})$

    Compute $d_t = \rho(S(\boldsymbol{x}^0), S(\boldsymbol{x}^{(t)}))$

end for

Order distances $d_{(1)} \leq d_{(2)} \leq \ldots \leq d_{(N_{ref})}$

return the values $\theta^{(t)}$ associated with the k smallest distances

---

Example 7. Getting back to the Student's t setting of Example 6, the p averages $\bar{x}_{n_i}$ contain information about the parameters μ and τ, but also exhibit variability that is not relevant to the approximation of the posterior probability. It thus makes sense to explore

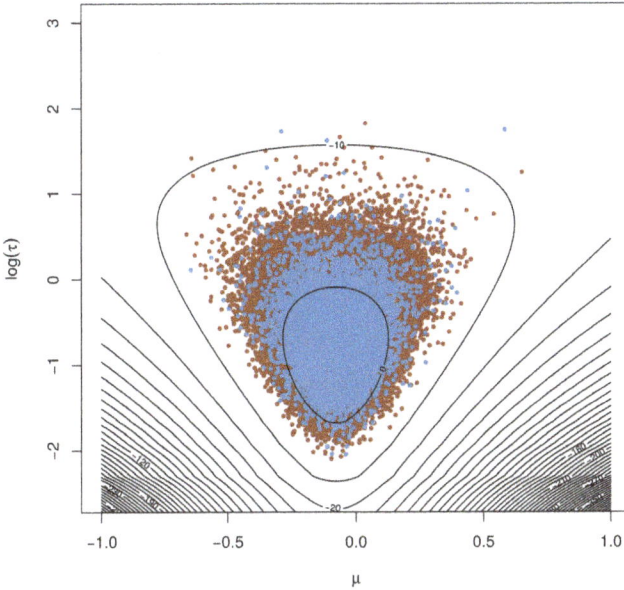

Fig. 6. Same legend as Fig. 5, representing the ABC posterior sample *(in blue)* along with the reference table *(in brown)* and with the level sets of the pseudo-posterior density *(in log scale)*.

the impact of considering solely the summaries

$$S(\bar{x}_{n_1},\ldots,\bar{x}_{n_p}) = (\bar{\bar{x}}, s^2)$$

already used in the construction of the pseudo-posterior. Algorithm 3 then implies generating pseudo-samples and comparing the values of their summary statistic through a distance. A major issue often overlooked in ABC applications is that the distance needs to be scaled, i.e., the plain sum of squares

$$(\bar{\bar{x}}_1 - \bar{\bar{x}}_2)^2 + (s_1^2 - s_1^2)^2$$

is not appropriate because both components are commensurable. Instead, we suggest using a normalised version like

$$(\bar{\bar{x}}_1 - \bar{\bar{x}}_2)^2 / \mathrm{mad}(\bar{\bar{x}})^2 + (s_1^2 - s_1^2)^2 / \mathrm{mad}(s^2)^2$$

where the median absolute deviation (MAD)

$$\mathrm{mad}(S(x)) = \mathrm{median}\,|S(x) - \mathrm{median}(S(x))|$$

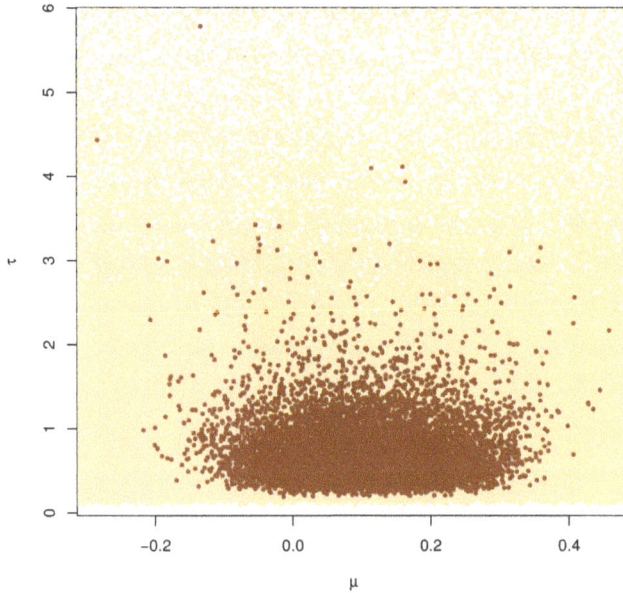

Fig. 7. Sample of $1,000$ simulations from Algorithm 3 when the data is made of 10 t averages with sample sizes $n_i = 21$ and when the 10^6 ABC simulations are taken from the prior. The summary statistic is made of the empirical mean and of the variance, while the distance is normalised by the MAD. *(Note: the true value of (μ, τ) is $(0, 1)$.)*

is estimated from the (prior) reference table simulated in the ABC algorithm. Running Algorithm 3 with this calibration produces an outcome summarised in Figures 7 and 8. The difference with Figure 4 is striking: while using the same prior, the outcome is not centred around the true value of the parameter in the former case while it is much more accurate in the second case.   ◄

The choice of the summary statistic is definitely essential to ensure ABC produces a reliable approximation to the true posterior distribution $\pi(\theta|x^0)$. A first important remark is that, at best, the outcome of Algorithm 3 will approximate simulation from $\pi(\theta|S(x^0))$. If the later strongly differs from $\pi(\theta|x^0)$, there is no way ABC can recover from this loss! Obviously, when $S(\cdot)$ is a sufficient statistic, there is no loss incurred but this is almost never the case, as exponential families very rarely call for the use of an ABC processing (see Grelaud et al. (2009) for an exception in the setting of the Ising model). A second remark is that, due to the nature of the ABC algorithm, namely the simulation of a huge reference table, followed by the selection of the "closest" parameters, several collections

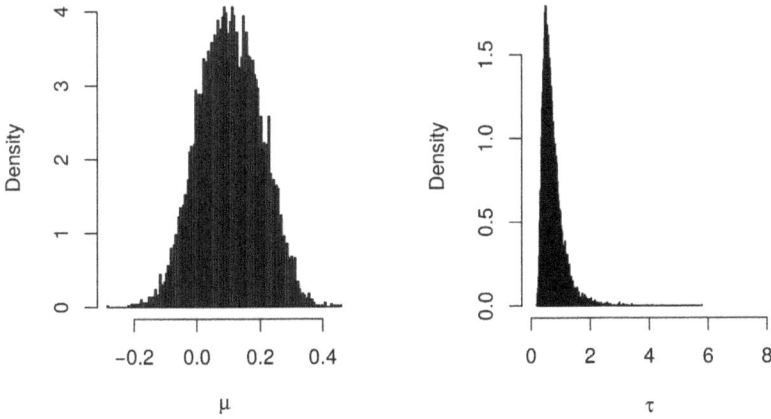

Fig. 8. Marginal histograms of a sample of $1,000$ simulations as in Fig. 7.

of summaries can be compared at a reasonable computational cost (assuming storing the entire pseudo-datasets a large number of time is feasible).

Example 8. Consider the most standard setting of a normal sample $x_1,\ldots,x_n \sim \mathcal{N}(\mu,\sigma^2)$ under a conjugate prior

$$\mu \sim \mathcal{N}(0,\tau^2), \sigma^{-2} \sim \mathcal{G}a(a,b).$$

If we decide to use the summary statistic $(\bar{x}_n, s_n^2)$, the (true) posterior will not change when compared with using the entire data, since this statistic is sufficient. On the other hand, if we use the pair $(\mathrm{med}(x_1,\ldots,x_n),\mathrm{mad}(x_1,\ldots,x_n))$, it is not sufficient and the (true) posterior will differ. Note that, in this second setting, this true posterior is not available as the joint distribution of the pair $(\mathrm{med}(x_1,\ldots,x_n),\mathrm{mad}(x_1,\ldots,x_n))$ is not available in closed form. Although there is no particular incentive to operate inference conditional on this pair, it provides a most simple illustration of a case when ABC must be used.

In this setting, in order to eliminate scaling effects due to some summaries varying more than others, we once more propose scaling by the mad statistics of those summaries,

$$\rho(S(\mathbf{x}^0), S(\mathbf{x}^{(t)})) = \sum_{i=1}^{2} |S_i(\mathbf{x}^0) - S_i(\mathbf{x}^{(t)}))|/\mathrm{mad}(S_i)$$

where $\mathrm{mad}(S_i)$ is thus based on the reference table.

When implementing ABC based on either of those pairs of summary statistics and a normal dataset of $5,000$ observations, Figure 9 shows that the outcome is identical! Furthermore, a comparison with the genuine output exhibits a significant difference, meaning that the impact of the tolerance is quite large in this case.   ◄

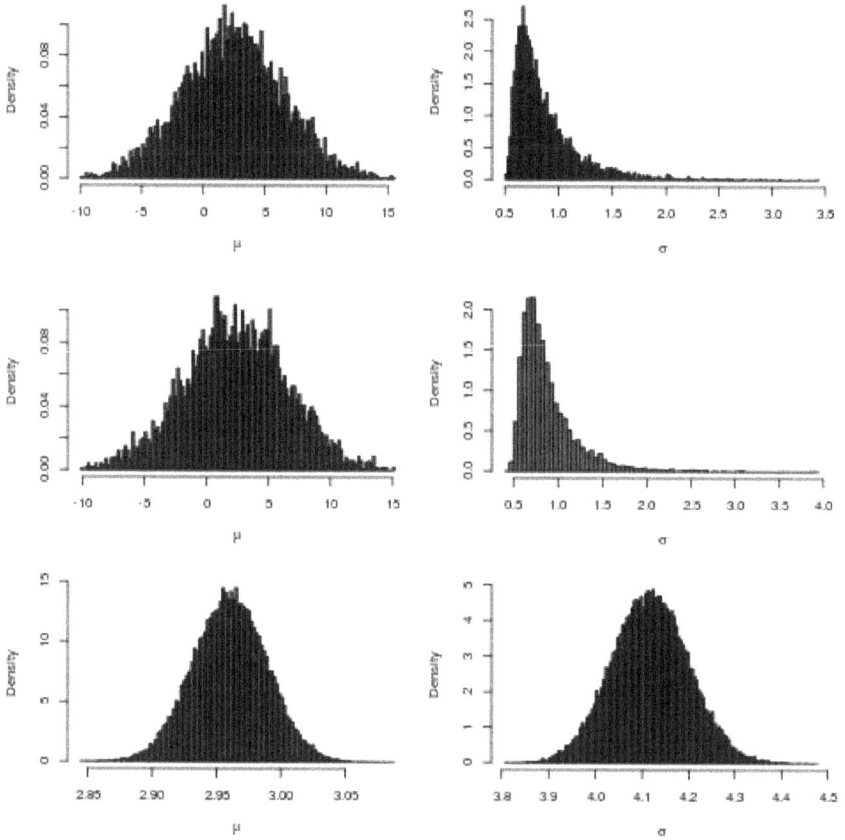

Fig. 9. Marginal histograms in μ and σ^2 *(left and right)* based on two ABC algorithms *(top and middle)* and on the (non-ABC) corresponding Gibbs sampler, for a sample of 5,000 normal $\mathcal{N}(2,4)$ observations, 10^6 ABC and Gibbs iterations, a subsampling rate of 5% for all algorithms and the use of the summary statistics $S(x_1,\ldots,x_n) = (\bar{x}_n, s_n^2)$ *(top)* and $S(x_1,\ldots,x_n) = (\mathrm{med}(x_1,\ldots,x_n), \mathrm{mad}(x_1,\ldots,x_n))$ *(middle)*.

2.5 Wikipedia entry

For a further introduction to ABC methods, I refer the reader to our earlier survey (Marin et al., 2011). I further recommend Sunnåker et al. (2013), the publication constituting the original version of the Wikipedia page on ABC (Wikipedia, 2014). the presentation made in that page is comprehensive and correct, rightly putting stress on the most important aspects of the method. The authors also properly warn about the need to assess assumptions behind and calibrations of the method. (Comments of both referees are included in the original paper, available on-line.)

Note that the ABC method was *not* introduced for conducting model choice, even though this implementation may currently constitute the most frequent application of the method, and the derivation of ABC model choice techniques appeared rather recently (Grelaud et al., 2009; Toni et al., 2009). In almost every setting where ABC is used, there is no non-trivial sufficient summary statistic. Relying on an insufficient statistic then implies a loss of statistical information, as discussed futher below, and I appreciate very much that the authors of the Wikipedia page advertise our warning (Robert et al., 2011) about the potential lack of validity when using an ABC approximation to a Bayes factor for model selection. I also like the notion of "quality control". And the pseudo-example is quite fine as an introduction, while it could be supplemented with the outcome resulting from a large n, to be compared with the true posterior distribution. The section "Pitfalls and remedies" is remarkable in that it details the necessary steps for validating an ABC implementation. A last comment is that the section on the non-zero tolerance could emphasise more strongly the fact that this tolerance ε *should not be zero*. (This recommendation may sound paradoxical, but from a practical perspective, $\varepsilon = 0$ can only be achieved with an infinite computing power.)

3   ABC Consistency

While ABC was first perceived with suspicion by the mainstream statistical community (as well as some population geneticists, see Templeton (2008, 2010); Beaumont et al. (2010); Berger et al. (2010), representations of the ABC posterior distribution as a true posterior distribution (Wilkinson, 2013) and of ABC as an auxiliary variable method (Wilkinson, 2013), as a non-parametric technique (Blum, 2010; Blum and François, 2010), connected with both indirect inference (Drovandi et al., 2011) and k-nearest neighbour estimation (Biau et al., 2015) helped to turn ABC into an acceptable component of Bayesian computational methods, albeit requiring caution and calibration (Wikipedia, 2014). The following entries cover some of the advances made in the statistical analysis of the method.

3.1   ABC as knn

Biau et al. (2015) made a significant contribution to the statistical foundations of ABC. It analyses the convergence properties of the ABC algorithm the way it is truly implemented (as in DIYABC (Cornuet et al., 2008) for instance), i.e. with a tolerance bound ε that is determined as a quantile of the simulated distances as in Algorithm 3, say the 10% or the 1% quantile. This means in particular that the interpretation of ε as a non-parametric density estimation bandwidth, while interesting and prevalent in the literature (see, e.g., Blum (2010) and Fearnhead and Prangle (2012)), is only an approximation of the actual practice.

The focus of Biau et al. (2015) is on the mathematical foundations of this practice, an advance obtained by (re)analysing ABC as a k-nearest neighbour (knn) method. Using generic knn results, they derive a consistency property for the ABC algorithm by imposing some constraints upon the rate of decrease of the quantile k as a function of n. More specifically, provided

$$k_N / \log \log N \longrightarrow \infty \quad \text{and} \quad k_N / N \longrightarrow 0$$

when $N \to \infty$, for almost all s_0 (with respect to the distribution of $S(Y)$), with probability 1, convergence occurs, i.e.

$$\frac{1}{k_N} \sum_{j=1}^{k_N} \varphi(\theta_j) \longrightarrow \mathbb{E}[\varphi(\theta_j)|S = s_0]$$

(The setting is restricted to the use of sufficient statistics or, equivalently, to a distance over the whole sample. The issue of summary statistics is not addressed by the paper.) The paper also contains a rigorous proof of the convergence of ABC when the tolerance ε goes to zero. The mean integrated square error consistency of the conditional kernel density estimate is established for a generic kernel (under usual assumptions). Further assumptions (both on the target and on the kernel) allow the authors to obtain precise convergence rates (as a power of the sample size), derived from classical k-nearest neighbour regression, like

$$k_N \approx N^{(p+4)/(m+p+4)}$$

in dimensions m larger than 4 (where N is the simulation size). The paper is theoretical and highly mathematical (with 25 pages of proofs!), but this work clearly constitutes a major reference for the justification of ABC. The authors also mention future work in that direction: I would suggest they consider the case of the insufficient summary statistics from this knn perspective.

3.2   Convergence Rates

Dean et al. (2014) addresses ABC consistency in the special setting of hidden Markov models (HMM). It relates to Fearnhead and Prangle (2012) discussed below in that those authors also establish ABC consistency for the noisy ABC, given in Algorithm 4, where $\mathfrak{h}(\cdot)$ is a kernel bounded by one (as for instance the unnormalised normal density).

Algorithm 4 ABC (noisy version)

    Compute $S(x^0)$ and generate $\tilde{S}^0 \sim \mathfrak{h}(\{s - S(x^0)\}/\varepsilon)$
    for $t = 1$ to N **do**
      repeat
        Generate θ^* from the prior $\pi(\cdot)$
        Generate x^* from the model $f(\cdot|\theta^*)$
        Accept θ^* with probability $\mathfrak{h}(\{\tilde{S}^0 - S(x)\}/\varepsilon)$
      until acceptance
    end for
    return N accepted values of θ^*

The authors construct an ABC scheme such that the ABC simulated sequence remains an HMM, the conditional distribution of the observables given the latent Markov chain being modified by the ABC acceptance ball. This means that conducting maximum likelihood (or Bayesian) estimation based on the ABC sample is equivalent to exact inference under the perturbed HMM scheme. In this sense, this equivalence bridges with Wilkinson (2013) and Fearnhead and Prangle (2012) perspectives on "exact ABC". While the paper provides asymptotic bias for a fixed value of the tolerance ε, it also proves that an arbitrary accuracy can be attained with enough data and a small enough ε. The authors of the paper show in addition (as in Fearnhead's and Prangle's) that an ABC inference based on noisy observations $y_1 + \varepsilon z_1, \ldots, y_n + \varepsilon z_n$ with the same tolerance ε, is equivalent to a regular inference based on the original data $y_1, \ldots, y_n$, hence the asymptotic consistence of Algorithm 4. Furthermore, the asymptotic variance of the ABC version is proved to always be greater than the asymptotic variance of the standard maximum likelihood estimator, and decreasing as ε^2. The paper also contains an illustration on an HMM with α-stable observables. (Of course, the restriction to summary statistics that preserve the HMM structure is paramount for the results in the paper to apply, hence preventing the use of truly summarising statistics that would not grow in dimension with the size of the HMM series.) In conclusion, a paper that validates noisy ABC without non-parametric arguments and which makes me appreciate even further the idea of noisy ABC: at first, I liked the concept but found the randomisation it involved rather counter-intuitive from a Bayesian perspective. Now, I perceive it as a duplication of the randomness in the data that brings the simulated model closer to the observed model.

Bornn et al. (2014) is a note on the convergence properties of ABC, when compared with acceptance-rejection or regular MCMC. Unsurprisingly, ABC does worse in both cases. What is central to this note is that ABC can be (re)interpreted as a pseudo-marginal method where the data comparison step acts like an unbiased estimator of the true ABC target (not of the original ABC target). From there, it is mostly an application of Andrieu and Vihola (2014) in this setup. The authors also argue that using a single pseudo-data simulation per parameter value is the optimal strategy (as compared with using several), when considering asymptotic variance. This makes sense in terms of simulating in a larger dimensional space but there may be a trade-off when considering the cost of producing those pseudo-datasets against the cost of producing a new parameter. There are a few (rare) cases where the datasets are much cheaper to produce.

Barber et al. (2013) is essentially theoretical and establishes the optimal rate of convergence of the mean square error—for approximating a posterior moment—at a rate of $2/(q+4)$, where q is the dimension of the summary statistic, associated with an optimal tolerance in $n^{-1/4}$, n being the sample size. I was first surprised at the role of the dimension of the summary statistic, but rationalised it as being the dimension where the non-parametric estimation takes place. There are obviously links with earlier convergence results found in the literature: for instance, Blum (2010) relates ABC to standard kernel density non-parametric estimation and finds a tolerance (bandwidth) of order $n^{-1/q+4}$ and an mean square error of order $2/(q+4)$ as well. Similarly, Biau et al. (2015) obtain precise convergence rates for ABC interpreted as a k-nearest-neighbour estimator (Section 3.1). And, as detailed in Section 4.1, Fearnhead and Prangle (2012) derive rates similar to Blum's with a tolerance of order $n^{-1/q+4}$ for the regular ABC and of order $n^{-1/q+2}$ for the noisy ABC.

3.3 Checking ABC Convergence

Prangle et al. (2013) is a paper on diagnostics for ABC validation via coverage diagnostics. Getting valid approximation diagnostics for ABC is clearly and badly needed. When simulation time is not an issue (!), the DIYABC (Cornuet et al., 2008) software does implement a limited coverage assessment by computing the type I error, i.e. by simulating data under the null model and evaluating the number of time it is rejected at the 5% level (see sections 2.11.3 and 3.8 in the documentation). The current paper builds on a similar perspective.

The core idea is that a (Bayesian) credible interval at a given credible level α should have a similar confidence level (at least asymptotically and even more for matching priors) and that simulating pseudo-data with a known parameter value allows for a Monte-Carlo evaluation of the credible interval "true" coverage, hence for a calibration of the tolerance. The delicate issue is about the generation of those "known" parameters. For instance, if the pair (θ, y) is generated from the joint distribution prior x likelihood, and if the credible region is also based on the true posterior, the average coverage is the nominal one. On the other hand, if the credible interval is based on a poor (ABC) approximation to the posterior, the average coverage should differ from the nominal one.

4   Summary Statistics, the ABC Conundrum

The main focus of the recent ABC literature has been on the selection and evaluation of summary statistics, including a Royal Statistical Society Read Paper (Fearnhead and Prangle, 2012) that set a reference and gave prospective developments in the discussion section. Reducing the data into a small dimension but sufficienlt informative statistics constitutes a fundamental difficulty when there is no non-trivial sufficient statistic and when the summary statistic is not already provided by the software (like DIYABC, Cornuet et al. (2008)) or imposed by experimenters in the field. This choice has to balance a loss of statistical information a gain in ABC precision, with little available on the amounts of error and information loss involved in the ABC substitution.

4.1 The Read Paper

Fearnhead and Prangle (2012) proposed an original approach to ABC, where ABC is considered from a purely inferential viewpoint and calibrated for estimation purposes. Fearnhead and Prangle do not follow the "traditional" perspective of looking at ABC as a converging approximation to the true posterior density. As Wilkinson (2013) (first posted in 2008), they take instead a randomised/noisy version of the summary statistic and derive a calibrated version of ABC, i.e. an algorithm that gives proper predictions, the drawback being that it is for the posterior given this randomised version of the summary statistic. The paper also contains an important result in the form of a consistency theorem that shows that noisy ABC is a convergent estimation method when the number of observations or datasets grows to infinity. The most interesting aspect in this switch of perspective is that the kernel $\mathfrak{h}$ used in the acceptance probability

$$\mathfrak{h}((s - s_{\mathrm{obs}})/h)$$

does not have to act as an estimate of the true sampling density, since it appears in the (randomised) pseudo-model. (Everything collapses to the true model when the bandwidth h goes to zero.) The Monte Carlo error is taken into account through the average acceptance probability, which collapses to zero when h goes to zero, therefore a suboptimal choice!

A form of tautology stems from the comparison of ABC posteriors via a loss function

$$(\theta_0 - \hat{\theta})^{\mathrm{T}} A (\theta_0 - \hat{\theta})$$

that ends up with the "best" asymptotic summary statistic being

$$\mathbb{E}[\theta | y_{\mathrm{obs}}].$$

This result indeed follows from the very choice of the loss function rather than from an intrinsic criterion. Using the posterior expectation as the summary statistic still makes sense, especially when the calibration constraint implies that the ABC approximation has the same posterior mean as the true (randomised) posterior. Unfortunately this result is parameterisation dependent and unlikely to be available in settings where ABC is necessary. In the semi-automatic implementation proposed by Fearnhead and Prangle (2012), the authors suggest to use a pilot run of ABC to approximate the above statistic. I wonder at the resulting cost since a simulation experiment must be repeated for each simulated dataset (or sufficient statistic). The simplification in the paper follows from a linear regression on the parameters, thus linking the approach with Beaumont et al. (2002).

Using the same evaluation via a posterior loss, the authors show that the "optimal" kernel is uniform over a region

$$x^{\mathrm{T}} A x < c$$

where c makes a ball of volume 1. A significant remark is that the error evaluated by Fearnhead and Prangle is

$$\mathrm{tr}(A\Sigma) + h^2 \mathbb{E}_{f_h}[x^{\mathrm{T}} A x] + \frac{C_0}{h^d}$$

which means that, due to the Monte Carlo error, the "optimal" value of h is not zero but akin to a non-parametric optimal speed in $2/2+d$. There should thus be a way to link this decision-theoretic approach with the one of Ratmann et al. (2009) since the latter take h to be part of the parameter vector.

As exposed in my discussion (Robert, 2012), I remain skeptical about the "optimality" resulting from the choice of summary statistic in the paper, partly because practice shows that proper approximation to genuine posterior distributions stems from using a (much) larger number of summary statistics than the dimension of the parameter (albeit unachievable at a given computing cost), partly because the validity of the approximation to the optimal summary statistics depends on the quality of the pilot run, and partly because there are some imprecisions in the mathematical derivation of the results (Robert, 2012). Furthermore, important inferential issues like model choice are not covered by this approach. But, nonetheless, the paper provides a way to construct default summary statistics that should come as a supplement to the summary statistic provided by the experts, or even as a substitute.

The paper is also connecting to the computing cost and stressing the relevance of the indirect inference literature (Gouriéroux et al., 1993). A clear strength of the paper remains with Section 4 which provides a major simulation experiment. My only criticism on this section would be about the absence of a phylogeny example that would relate to the models that launched ABC methods. This is less of a mainstream statistics example, but it would be highly convincing to those primary users of ABC.

4.2 Another Review

"What is very apparent from this study is that there is no single 'best' method of dimension reduction for ABC." M. Blum, M. Nunes, D. Prangle, and S. Sisson (2013)

Blum et al. (2013) offers a detailed review of dimension reduction methods in ABC, along with a comparison on three specific models. Given that, as put above, the choice of the vector of summary statistics is presumably the most important single step in an ABC algorithm and keeping in mind that selecting too large a vector is bound to fall victim of the dimension curse, this constitutes a reference for the ABC literature. Therein, the authors compare regression adjustments à la Beaumont et al. (2002), subset selection methods, as in Joyce and Marjoram (2008), and projection techniques, as in Fearnhead and Prangle (2012). They add to this impressive battery of methods the potential use of AIC and BIC. An argument for using AIC/BIC is that either provides indirect information about the approximation of $p(\theta|y)$ by $p(\theta|s(y))$, even though this does not seem obvious to me.

The paper also suggests a further regularisation of Beaumont et al. (2002) by ridge regression, although L_1 penalty à la Lasso would be more appropriate in my opinion for removing extraneous summary statistics. (I must acknowledge never being a big fan of ridge regression, esp. in the ad hoc version à la Hoerl and Kennard (1970), i.e. in a non-decision theoretic approach where the hyperparameter λ is derived from the data by cross-validation, since it then sounds like a poor man's version of Bayes' and Stein' estimators, just like BIC is a first order approximation to regular Bayes factors). Unsurprisingly, ridge regression does better than plain regression in the comparison experiment when there are many almost collinear summary statistics, but an alternative conclusion could be that regression analysis is not that appropriate with many summary statistics. Indeed, summary statistics are not quantities of interest but data summarising tools towards a better approximation of the posterior at a given computational cost. (I do not get the final comment about the relevance of summary statistics for MCMC or SMC algorithms: the criterion should be the best approximation of $p(\theta|y)$ which does not depend on the type of algorithm.)

5 ABC Model Choice

While ABC is a substitute for a proper—possibly MCMC based—Bayesian inference, and thus pertains to all aspects of Bayesian inference, including testing and model checking, the special issue of comparing models via ABC is highly delicate and concentrated most of the criticisms addressed against ABC (Templeton, 2008, 2010). The implementation of ABC model choice follows by treating the model index m as an extra parameter with an associated prior, as detailed in the following algorithm:

Algorithm 5 ABC (model choice)

---

  for $i = 1$ to N **do**
    repeat
      Generate $\mathfrak{m}$ from the prior $\pi(\mathcal{M} = m)$
      Generate θ_m from the prior $\pi_m(\theta_m)$
      Generate z from the model $f_m(z|\theta_m)$
    until $\rho\{S(z), S(y)\} \leq \varepsilon$
    Set $\mathfrak{m}^{(i)} = \mathfrak{m}$ and $\theta^{(i)} = \theta_m$
  end for
  return the values $\mathfrak{m}^{(i)}$ associated with the k smallest distances

---

Improvements upon returning raw model index frequencies as ABC estimates have been proposed in Fagundes et al. (2007), via a regression regularisation. In this approach, indices are processed as categorical variables in a formal multinomial regression, using for instance logistic regression. Rejection-based approaches as in Algorithm 5 were introduced in Cornuet et al. (2008), Grelaud et al. (2009) and Toni et al. (2009), in a Monte Carlo perspective simulating model indices as well as model parameters. Those versions are widely used by the population genetics community, as exemplified by Belle et al. (2008); Cornuet et al. (2010); Excoffier et al. (2009); Ghirotto et al. (2010); Guillemaud et al. (2009); Leuenberger and Wegmann (2010); Patin et al. (2009); Ramakrishnan and Hadly (2009); Verdu et al. (2009); Wegmann and Excoffier (2010). As described in the following sections, this adoption may be premature or over-optimistic, since caution and cross-checking are necessary to completely validate the output.

5.1   A Clear Lack of Confidence

In Robert et al. (2011), we came to the conclusion, shocking to us, that ABC approximations to posterior probabilities cannot be uniformly trusted. Approximating posterior probabilities by an ABC algorithm, i.e. by using the frequencies of acceptances of simulations from those models (assuming the use of a common summary statistic to define the distance to the observations) offer no general guarantees. Rather obviously (*ex post!*), we ended up with the limiting behaviour being ruled by a true Bayes factor, except it is the one based on the distributions of the summary statistics under both models.

At first, this did not sound a particularly novel and fundamental result, since all ABC approximations rely on the posterior distributions based on those summary statistics, rather than on the whole dataset. However, while this approximation only has consequences in terms of the precision of the inference for most inferential purposes, it induces a dramatic arbitrariness in the Bayes factor. To illustrate this arbitrariness, consider the case of using a sufficient statistic $S(x)$ for both models. Then, by the factorisation theorem, the true likelihoods factorise as

$$\ell_1(\theta_1|x) = g_1(x)p_1(\theta_1|S(x)) \quad \text{and} \quad \ell_2(\theta_2|x) = g_2(x)p_2(\theta_2|S(x))$$

resulting in a true Bayes factor equal to

$$B_{12}(x) = \frac{g_1(x)}{g_2(x)} B_{12}^S(x), \tag{5.1}$$

where the last term $B_{12}^S(x)$ is the limiting ABC Bayes factor. Therefore, in the favourable case of the existence of a sufficient statistic, using only the sufficient statistic induces a difference in the result that fails to converge with the number of observations or simulations. On the opposite, it may diverge one way or another as the number of observations increases. Again, this is in the favourable case of sufficiency. In the realistic setting of using summary statistics, things deteriorate further! This practical situation indeed implies a wider loss of information compared with the exact inferential approach, hence a wider discrepancy between the exact Bayes factor and the quantity produced by an ABC approximation. It thus appeared to us as an urgent duty to warn the community about the dangers of this approximation, especially when considering the rapidly increasing number of applications using ABC for conducting model choice and hypothesis testing. Furthermore, we unfortunately did not see at the time an immediate and generic alternative for the approximation of Bayes factor.

The paper stresses what I consider a fundamental or even foundational distinction between ABC point (and confidence) estimation and ABC model choice, namely that the problem was at another level for Bayesian model choice (using posterior probabilities). When doing point estimation with insufficient summary statistics, the information content is poorer, but unless one uses very degraded summary statistics, inference is converging. The posterior distribution is still different from the true posterior in this case but, at least, gathering more observations brings more information about the parameter (and convergence when the number of observations goes to infinity). For model choice, this is not guaranteed if we use summary statistics that are not inter-model sufficient, as shown by the Poisson and normal examples. Furthermore, except for very specific cases such as Gibbs random fields (Grelaud et al., 2009), it is almost always impossible to derive inter-model sufficient statistics, beyond the raw sample.

Example 9. Another example is described in the introduction of the "sequel" by Marin et al. (2014), to be discussed below. The setting is one of a comparison between a normal $y \sim \mathcal{N}(\theta_1, 1)$ model and a double exponential $y \sim \mathcal{L}(\theta_2, 1/\sqrt{2})$ model.[6] The summary statistics used in the corresponding ABC algorithm are the sample mean, the sample median and the sample variance. Figure 10 exhibits the absence of discrimination between the two models, since the posterior probability of the normal model converges to a central value around 0.5-0.6 when the sample size grows, irrelevant of the true model behind the simulated datasets! ◀

The paper includes a realistic population genetic illustration, where two scenarios including three populations were compared, two populations having diverged 100 generations ago and the third one resulting of a recent admixture between the first two populations (scenario 1) or simply diverging from population 1 (scenario 2) at the same time of 5 generations in the past. In scenario 1, the admixture rate is 0.7 from population 1. Pseudo observed datasets (100) of the same size as in experiment 1 (15 diploid individuals per

---

[6]The double exponential distribution is also called the Laplace distribution, hence the notation $\mathcal{L}(\theta_2, 1/\sqrt{2})$, with mean θ_2 and variance one.

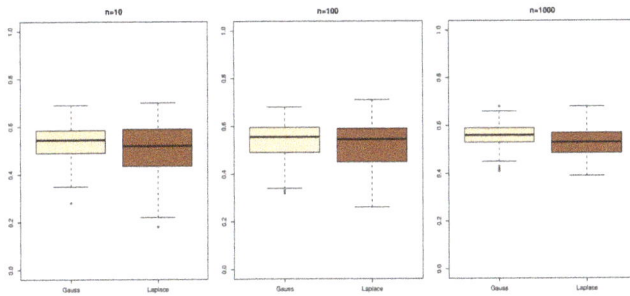

Fig. 10. Comparison of the range of the ABC posterior probability that data is from a normal model *(and not from a Laplace model)* with unknown mean θ when the data is made of $n = 10, 100, 1000$ observations *(left, center, right, resp.)* either from a Gaussian *(lighter)* or Laplace distribution *(darker)* and when the ABC summary statistic is made of the empirical mean, median, and variance. The ABC algorithm generates 10^4 simulations (5,000 for each model) from the prior $\theta \sim \mathcal{N}(0,4)$ and selects the tolerance ε as the 1% distance quantile over those simulations. *(Source: Marin et al. (2014).)*

population, 5 independent microsatellite loci) have been generated assuming an effective population size of 1000 and mutation rates of 0.0005. There are six parameters (provided with the corresponding priors): admixture rate (U[0.1,0.9]), three effective population sizes (U[200,2000]), the time of admixture/second divergence (U[1,10]) and the time of the first divergence (U[50,500]). Although this is rather costly in computing time, the posterior probability can nonetheless be estimated by importance sampling, based on 1000 parameter values and 1000 trees per parameter value, based on the modules of Stephens and Donnelly (2000). The ABC approximation is obtained from DIYABC (Cornuet et al., 2008), using a reference sample of two million parameters and 24 summary statistics. The result of this experiment is shown above, with a clear divergence in the numerical values despite stability in both approximations. Taking the importance sampling approximation as the reference value, the error rates in using the ABC approximation to choose between scenarios 1 and 2 are 14.5% and 12.5% (under scenarios 1 and 2), respectively. Although a simpler experiment with a single parameter and the same 24 summary statistics shows a reasonable agreement between both approximations, this result comes an additional support to our warning about a blind use of ABC for model selection. The corresponding simulation experiment was quite intense, as, with 50 markers and 100 individuals, the product likelihood suffers from such an enormous variability that 100,000 particles and 100 trees per locus have trouble to address (despite a huge computing cost of more than 12 days on a powerful cluster).

A quite related if less pessimistic paper is Didelot et al. (2011), also concerned with the limiting behaviour for the ratio (5.1) Indeed, the authors reach the opposite conclusion from ours, namely that the problem can be solved by a sufficiency argument. Their point is that, when comparing models within exponential families (which is the natural realm for sufficient statistics), it is always possible to build an encompassing model with a sufficient statistic that remains sufficient across models. This construction is correct from a mathematical perspective, as seen for instance in the Poisson versus geometric example we first mentioned in Grelaud et al. (2009): adding

$$\prod_{i=1}^{n} x_i!$$

to the sum of the observables into a large sufficient statistic produces a ratio g_1/g_2 that is equal to 1, hence avoids any discrepancy.

Nonetheless, we do not think this encompassing property has a direct impact on the performances of ABC model choice. In practice, complex models do not enjoy sufficient statistics (if only because the overwhelming majority of them are not exponential families, with the notable exception of Gibbs random fields where the above agreement graph is derived). There is therefore a strict loss of information in using ABC model choice, due to the call both to insufficient statistics and to non-zero tolerances. Looking at what happens in the limiting case when one is relying on a common sufficient statistic is a formal study that brings light on the potentially huge discrepancy between the ABC-based Bayes factor and the true Bayes factor. This is why we consider that finding a solution in this formal case—while a valuable extension of the Gibbs random fields case—does not directly help towards the understanding of the discrepancy found in non-exponential complex models.

5.2   Validating Summaries for ABC Model Choice

Our answer to the (well-received) above warning is provided in Marin et al. (2014), which deals with the evaluation of summary statistics for Bayesian model choice. Even though the idea of separating the mean behaviour of the statistics under both model came rather early, establishing a complete theoretical framework that validated this intuition took quite a while and the assumptions changed a few times around the summer. The simulations associated with the paper were straightforward in that (a) the setup had been suggested to us by a referee (Robert et al., 2011): as detailed in Example 9, they consist in comparing normal and Laplace distributions with different summary statistics (inc. the *median absolute deviation*, which is the median of the absolute deviation from the median, $\text{med}(|x - \text{med}(x)|)$); (b) the theoretical results told us what to look for; and (c) they did very clearly exhibit the consistency and inconsistency of the Bayes factor/posterior probability predicted by the theory. Both boxplots shown on Figures 10 and 11 exhibit this agreement: when using (empirical) mean, median, and variance to compare normal and Laplace models, the posterior probabilities do not select the "true" model but instead aggregate near a fixed value. Hence ABC based on those summary statistics is not discriminative. When using instead the median absolute deviation as summary statistic, the posterior probabilities concentrate near one or zero depending on whether or not the normal model is the true model. Hence, this summary statistic is highly discriminant for the comparison of the two models. From an ABC perspective, this means that using the median absolute deviation is then satisfactory, as opposed to the above three statistics.

The above example illustrates very clearly the major result of this paper, namely that the mean behaviour of the summary statistic $S(y)$ under both models under comparison is

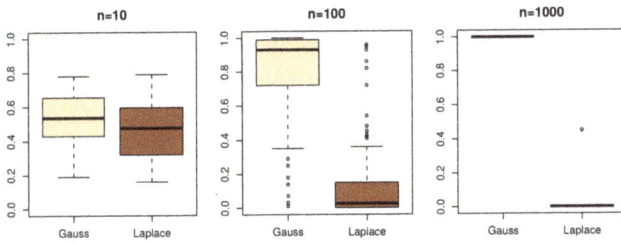

Fig. 11. Comparison of the distributions of the posterior probabilities that the data is from a normal model *(as opposed to a Laplace model)* with unknown mean θ when the data is made of $n = 10, 100, 1000$ observations *(left, center, right, resp.)* either from a Gaussian or Laplace distribution with mean equal to zero and when the summary statistic in the ABC algorithm is the median absolute deviation. The ABC algorithm uses a reference table of 10^4 simulations (5,000 for each model) from the prior $\theta \sim \mathcal{N}(0,4)$ and selects the tolerance ε as the 1% distance quantile over those simulations.

fundamental for the convergence of the Bayes factor, i.e. of the Bayesian model choice based on $S(y)$. This result, described in the next section, thus brings an almost definitive answer to the question raised in Robert et al. (2011) about the validation of ABC model choice.

More precisely, Marin et al. (2014) states that, under some "heavy-duty" Bayesian asymptotics assumptions, (a) if the "true" mean of the summary statistic can be recovered for both models under comparison, then the Bayes factor is of order

$$O\left(n^{-(d_1 - d_2)/2}\right),$$

where d_i is the intrinsic dimension of the parameters driving the summary statistic in model $i = 1, 2$, irrespective of which model is true. (Precisely, the dimensions d_i are the dimensions of the asymptotic mean of the summary statistic under both models.) Therefore, the Bayes factor always asymptotically selects the model having the smallest effective dimension and cannot be consistent. (b) if, instead, the "true" mean of the summary statistic cannot be represented in the other model, then the Bayes factor is consistent. This means that, somehow, the best statistics to be used in an ABC approximation to a Bayes factor are ancillary statistics with different mean values under both models. Else, the summary statistic must have enough components to prohibit a parameter under the "wrong" model to meet the "true" mean of the summary statistic.

One of the referee's comments on the paper was that maybe Bayes factors were not appropriate for conducting model choice, thus making the whole derivation irrelevant. This is a possible perspective but it can be objected that Bayes factors and posterior probabilities are used in conjunction with ABC in dozens of genetic papers. Further arguments are provided in the various replies to both of Templeton's radical criticisms (Templeton, 2008, 2010). That more empirical and model-based assessments also are available is quite correct, as demonstrated in the multicriterion approach of Ratmann et al. (2009). This is simply another approach, not followed by most geneticists so far.

Another criticism was that the paper is quite theoretical and the mathematical assumptions required to obtain the convergence theorems are rather overwhelming. Meaning that in practical cases they cannot truly be checked. However, I think we can eventually address those concerns for two distinct reasons: first, the paper comes as a third step in a series of papers where we first identified a sufficiency property, then realised that this property was actually quite a rare occurrence, and finally made a theoretical advance as to when is a summary statistic enough (i.e. "sufficient" in the standard sense of the term!) to conduct model choice, with a clear answer that the mean ranges of the summary statistic under each model could not intersect. Second, my own personal view is that those assumptions needed for convergence are not of the highest importance for statistical practice (even though they are needed in the paper!) and thus that, from a methodological point of view, only the conclusion should be taken into account. It is then rather straightforward to come up with (quick-and-dirty) simulation devices to check whether a summary statistic behaves differently under both models, taking advantage of the reference table already available (instead of having to run Monte Carlo experiments with ABC basis). The final version of the paper (Marin et al., 2014) includes a χ^2 check about the relevance of a given summary statistic.

At last, we could not answer in depth a query about the different speeds of convergence of the posterior probabilities under the Gaussian and Laplace distributions. This was a most interesting question in that the marginal likelihoods do indeed seem to converge at different speeds. However, the only precise information we can derive from our result (Theorem 1) is when the Bayes factor is not consistent. Otherwise, we only have a lower bound on its speed of convergence (under the correct model). Getting precise speeds in this case sounds beyond our reach...

5.3   Towards Estimating Posterior Probabilities

Stoehr et al. (2014) attack the recurrent problem of selecting summary statistics for ABC in a hidden Markov random field, where is no fixed dimension sufficient statistics. The paper provides a very broad overview of the issues and difficulties related with ABC model choice, which has been the focus of some advanced research only for a few years. Most interestingly, the authors define a novel, local, and somewhat Bayesian misclassification rate, an error that is conditional on the observed value and derived from the ABC reference table. It is the posterior predictive error rate

$$\mathbb{P}^{\text{ABC}}(\hat{m}(y^{\text{obs}}) \neq m | S(y^{\text{obs}}))$$

integrating in both the model index m and the corresponding random variable Y (and the hidden intermediary parameter) given the observation. Or rather given the transform of the observation by the summary statistic S. The authors even go further to define the error rate of a classification rule based on a first (collection of) statistic, conditional on a second (collection of) statistic (see Definition 1). A notion rather delicate to validate on a fully Bayesian basis. And they advocate the substitution of the unreliable (estimates of the) posterior probabilities by this local error rate, estimated by traditional non-parametric kernel methods. Methods that are calibrated by cross-validation. Given a reference summary statistic, this perspective leads (at least in theory) to select the optimal summary statistic as the one leading to the minimal local error rate. Besides its application to hidden Markov random fields, which is of interest per se, this paper thus opens a new vista on calibrating ABC methods and evaluating their true performances conditional on the actual data. The advocated abandonment of the estimation of all posterior probabilities could almost justify the denomination of a paradigm shift. This is also the approach advocated in Pudlo et al. (2015).

However, the above posterior predictive error rate is the conditional expected value of a misclassification loss when conditioning on the data (or more precisely some summaries of the data) being what it is. Hence, when integrating this conditional error over the marginal distribution of the summaries of the data, we recover the misclassification error integrated over the whole prior space. This quantity differs from the posterior (predictive) error rate computed in an initial version of Pudlo et al. (2015), which involves an expectation over the predictive distribution given the observed data and thus, a second integral over the data space. As a consequence, the conditional error rates of Stoehr et al. (2014) is on the same ground as the posterior probabilities.

Pudlo et al. (2015) offers the central arguments that (a) using random forests is a good tool for choosing the most appropriate model, (b) evaluating the posterior misclassification error is available via standard ABC arguments, and (c) estimating the posterior probability of the selected model is possible via further random forests. The call to the machine-learning tool of a random forest (Breiman, 2001), traditionally used in classification, may sound at first at odds with a Bayesian approach, but it becomes completely justified once one sets the learning set as generated from the prior predictive distribution. A random forest is then a randomised version of a non-parametric predictor of the model index given the data. Note that Pham et al. (2014) also use random forests for ABC parameter estimation.

Let us briefly recall that a random forest aggregates classification trees, CART, (Breiman et al., 1984) by introducing for each tree a randomisation step represented in Algorithm 6 and consisting in bootstrapping the original sample and subsampling the summary statistics at each node of the tree. A CART is a binary classification tree that partitions the covariate space towards a prediction of the class index. Each node of this tree consists in a rule of the form $S_j < t_j$, where S_j is one of the covariates and t_j is chosen towards minimising an heterogeneity index (Hastie et al., 2009). In ABC model choice, a CART tree is calibrated from the reference table and it returns a model index for the observed summary statistic s^{obs}, following a path according to these binary rules.

Algorithm 6 Randomised CART

---

  start the tree with a single root
  repeat
    pick a non-homogeneous tip v such that $Q(v) \neq 1$
    attach to v two daughter nodes v_1 and v_2
    draw a random subset of covariates of size n_{try}
    for all covariates X_j in the random subset **do**
      find the threshold t_j in the rule $S_j < t_j$ that minimises $N(v_1)Q(v_1) + N(v_2)Q(v_2)$
    end for
    find the rule $S_j < t_j$ that minimises $N(v_1)Q(v_1) + N(v_2)Q(v_2)$ in j **and set** this best rule to node v
  until all tips v are homogeneous ($Q(v) = 0$)
  set the labels of all tips

Reproduced with permission of the authors from Pudlo et al. (2015).

Pudlo et al. (2015) then selects the most likely model among a collection of models, based on a random forest classifier made of several hundreds CARTs as illustrated below, as a majority vote decision, i.e., the most fequently allocated model among the trees.

Algorithm 7 RF for classification

---

  for $b = 1$ **to** B **do**
    draw a bootstrap sub-sample Z^* of size N_{boot} from the training data
    grow a tree T_b trained on Z^* with Algorithm 6
  end for
  output the ensemble of trees $\{T_b, b = 1 \ldots B\}$

Reproduced with permission of the authors from Pudlo et al. (2015).

A first approach envisioned random forests as a mere filter applied to a large set of summary statistics in order to produce a relevant subset of significant statistics, with the additional appeal of an associated distance between datasets induced by the forest itself. However, we later realised that (a) further ABC steps were counterproductive, once the model was selected by the random forest; (b) including more summary statistics was always beneficial to the performances of the forest; and (c) the connections between (i) the true posterior probability of a model, (ii) the ABC version of this probability, (iii) the random forest frequency approximating the above, were at best very loose. While the random forest approach offers the advantage of incorporating all available summary statistics and not imposing a preliminary selection among those, it obviously weights the most discriminating ones more heavily. For instance, in Pudlo et al. (2015), the linear discriminant analysis (LDA) components are among the most often used. Experimentt in Pudlo et al. (2015) show that the frequencies of the various models produced by Algorithm 5 are however not directly related with their posterior probabilities.

Exploiting the approach of Stoehr et al. (2014), Pudlo et al. (2015) still managed to produce a reliable estimate of those. Indeed, the posterior expected error associated with the 0–1 loss (Robert, 2001)

$$\mathbb{I}(\hat{m}(s^{obs}) \neq m) \tag{5.2}$$

where $\hat{m}(s^{obs})$ is the model selection procedure, can be shown to satisfy (Pudlo et al., 2015)

$$\mathbb{E}[\mathbb{I}(\hat{m}(s^{obs}) \neq m)|s^{obs}] = 1 - \mathbb{P}[m = \hat{m}(s^{obs})|s^{obs}].$$

This expected loss is thus the complement to the posterior probability that the true model is the MAP. While it is not directly available, it can be estimated from the reference table as a regression of m or more exactly $\mathbb{I}(\hat{m}(s) \neq m)$ over s^{obs}. A natural solution in this context is to use another random forest, producing a function $\rho(s)$ that estimates $\mathbb{P}[m \neq \hat{m}(s)|s]$ and to apply this function to the actual observations to deduce $1 - \rho(s^{obs})$ as an estimate of $\mathbb{P}[m = \hat{m}(s^{obs})|s^{obs}]$.

6 Conclusion

This chapter reflects upon the diversity and the many directions of progress in the field of ABC research. The overall take-home message is that the on-going research in this area has led both to consider ABC as part of the statistical toolbox and to envision different approaches to statistical modelling, where a complete representation of the whole world is no always feasible. Following the evolution of ABC in the past fifteen years we have thus moved from constructing approximate methods to accepting working with approximate models, a positive move in my opinion. I have not covered here the many existing packages and codes that have been developped in different languages for the last decades, reflecting this dynamic. (See for instance the wide ranging abctools R package of Nunes and Prangle (2015).)

References

Andrieu, C. and Roberts, G. (2009). The pseudo-marginal approach for efficient Monte Carlo computations. *Ann. Statist.*, 37(2):697–725.

Andrieu, C. and Vihola, M. (2014). Establishing some order amongst exact approximations of MCMCs. *ArXiv e-prints*.

Arratia, R. and DeSalvo, S. (2012). On the Random Sampling of Pairs, with Pedestrian examples. *ArXiv e-prints*.

Barber, S., Voss, J., and Webster, M. (2013). The Rate of Convergence for Approximate Bayesian Computation. *ArXiv e-prints*.

Beaumont, M. (2010). Approximate Bayesian computation in evolution and ecology. *Annual Review of Ecology, Evolution, and Systematics*, 41:379–406.

Beaumont, M., Nielsen, R., Robert, C., Hey, J., Gaggiotti, O., Knowles, L., Estoup, A., Mahesh, P., Coranders, J., Hickerson, M., Sisson, S., Fagundes, N., Chikhi, L., Beerli, P., Vitalis, R., Cornuet, J.-M., Huelsenbeck, J., Foll, M., Yang, Z., Rousset, F., Balding,

D., and Excoffier, L. (2010). In defense of model-based inference in phylogeography. *Molecular Ecology*, 19(3):436–446.

Beaumont, M., Zhang, W., and Balding, D. (2002). Approximate Bayesian computation in population genetics. *Genetics*, 162:2025–2035.

Belle, E., Benazzo, A., Ghirotto, S., Colonna, V., and Barbujani, G. (2008). Comparing models on the genealogical relationships among Neandertal, Cro-Magnoid and modern Europeans by serial coalescent simulations. *Heredity*, 102(3):218–225.

Berger, J., Fienberg, S., Raftery, A., and Robert, C. (2010). Incoherent phylogeographic inference. *Proc. National Academy Sciences*, 107(41):E57.

Biau, G., Cérou, F., and Guyader, A. (2015). New insights into Approximate Bayesian Computation. *Annales de l'IHP (Probability and Statistics)*, 51:376–403.

Blum, M. (2010). Approximate Bayesian Computation: a non-parametric perspective. *J. American Statist. Assoc.*, 105(491):1178–1187.

Blum, M. and François, O. (2010). Non-linear regression models for approximate Bayesian computation. *Statist. Comput.*, 20:63–73.

Blum, M. G. B., Nunes, M. A., Prangle, D., and Sisson, S. A. (2013). A comparative review of dimension reduction methods in Approximate Bayesian computation. *Statistical Science*, 28(2):189–208.

Bornn, L., Pillai, N., Smith, A., and Woodard, D. (2014). A Pseudo-Marginal Perspective on the ABC Algorithm. *ArXiv e-prints*.

Breiman, L. (2001). Random forests. *Machine Learning*, 45(1):5–32.

Breiman, L., Friedman, J., Stone, C. J., and Olshen, R. A. (1984). *Classification and regression trees*. CRC press.

Cornuet, J.-M., Ravigné, V., and Estoup, A. (2010). Inference on population history and model checking using DNA sequence and microsatellite data with the software DIYABC (v1.0). *BMC Bioinformatics*, 11:401.

Cornuet, J.-M., Santos, F., Beaumont, M., Robert, C., Marin, J.-M., Balding, D., Guillemaud, T., and Estoup, A. (2008). Inferring population history with DIYABC: a user-friendly approach to Approximate Bayesian Computation. *Bioinformatics*, 24(23):2713–2719.

Dean, T., Singh, S., Jasra, A., and Peters, G. (2014). Parameter inference for hidden Markov models with intractable likelihoods. *Scand. J. Statist.* (to appear).

Dempster, A., Laird, N., and Rubin, D. (1977). Maximum likelihood from incomplete data via the EM algorithm (with discussion). *J. Royal Statist. Society Series B*, 39:1–38.

Didelot, X., Everitt, R., Johansen, A., and Lawson, D. (2011). Likelihood-free estimation of model evidence. *Bayesian Analysis*, 6:48–76.

Diggle, P. and Gratton, R. (1984). Monte Carlo methods of inference for implicit statistical models. *J. Royal Statist. Society Series B*, 46:193–227.

Doucet, A., de Freitas, N., and Gordon, N. (1999). *Sequential MCMC in Practice*. Springer-Verlag.

Drovandi, C., Pettitt, A., and Fddy, M. (2011). Approximate Bayesian computation using indirect inference. *J. Royal Statist. Society Series A*, 60(3):503–524.

Excoffier, C., Leuenberger, D., and Wegmann, L. (2009). Bayesian computation and model selection in population genetics. arXiv:0901.2231.

Fagundes, N., Ray, N., Beaumont, M., Neuenschwander, S., Salzano, F., Bonatto, S., and Excoffier, L. (2007). Statistical evaluation of alternative models of human evolution. *Proceedings of the National Academy of Sciences*, 104(45):17614–17619.

Fearnhead, P. and Prangle, D. (2012). Constructing summary statistics for Approximate Bayesian Computation: semi-automatic Approximate Bayesian Computation. *Journal of the Royal Statistical Society: Series B (Statistical Methodology)*, 74(3):419–474. (With discussion.).

Feller, W. (1970). *An Introduction to Probability Theory and its Applications*, volume 1. John Wiley, New York.

Frazier, D. T., Martin, G. M., Robert, C. P., and Rousseau, J. (2018). Asymptotic properties of approximate bayesian computation. *Biometrika*. (To appear).

Gelman, A., Carlin, J., Stern, H., Dunson, D., Vehtari, A., and Rubin, D. (2013). *Bayesian Data Analysis*. Chapman and Hall, New York, New York, third edition.

Ghirotto, S., Mona, S., Benazzo, A., Paparazzo, F., Caramelli, D., and Barbujani, G. (2010). Inferring genealogical processes from patterns of bronze-age and modern DNA variation in Sardinia. *Mol. Biol. Evol.*, 27(4):875–886.

Golightly, A. and Wilkinson, D. (2011). Bayesian parameter inference for stochastic biochemical network models using particle MCMC. *Interface Focus*, 1(6):807–820.

Gouriéroux, C., Monfort, A., and Renault, E. (1993). Indirect inference. *J. Applied Econometrics*, 8:85–118.

Grelaud, A., Marin, J.-M., Robert, C., Rodolphe, F., and Tally, F. (2009). Likelihood-free methods for model choice in Gibbs random fields. *Bayesian Analysis*, 3(2):427–442.

Guillemaud, T., Beaumont, M., Ciosi, M., Cornuet, J.-M., and Estoup, A. (2009). Inferring introduction routes of invasive species using approximate Bayesian computation on microsatellite data. *Heredity*, 104(1):88–99.

Hastie, T., Tibshirani, R., and Friedman, J. (2001). *The Elements of Statistical Learning*. Springer-Verlag, New York.

Hastie, T., Tibshirani, R., and Friedman, J. (2009). *The elements of statistical learning. Data mining, inference, and prediction.* Springer Series in Statistics. Springer-Verlag, New York, 2 edition.

Hoerl, A. and Kennard, R. (1970). Ridge regression: biased estimators for non-orthogonal problems. *Technometrics*, 12:55–67.

Jaakkola, T. and Jordan, M. (2000). Bayesian parameter estimation via variational methods. *Statistics and Computing*, 10:25–37.

Joyce, P. and Marjoram, P. (2008). Approximately sufficient statistics and Bayesian computation. *Statistical Applications in Genetics and Molecular Biology*, 7(1):article 26.

Leuenberger, C. and Wegmann, D. (2010). Bayesian computation and model selection without likelihoods. *Genetics*, 184(1):243–252.

Li, W. and Fearnhead, P. (2018). On the asymptotic efficiency of approximate bayesian computation estimators. *Biometrika*. (To appear.).

Lunn, D., Thomas, A., Best, N., and Spiegelhalter, D. (2010). *The BUGS Book: A Practical Introduction to Bayesian Analysis*. Chapman & Hall/CRC Press.

Marin, J., Pillai, N., Robert, C., and Rousseau, J. (2014). Relevant statistics for Bayesian model choice. *J. Royal Statist. Society Series B*, 76(5):833–859.

Marin, J., Pudlo, P., Robert, C., and Ryder, R. (2011). Approximate Bayesian computational methods. *Statistics and Computing*, 21(2):279–291.

Nunes, M. A. and Prangle, D. (2015). abctools: An R package for tuning Approximate Bayesian Computation analyses. Forthcoming.

Patin, E., Laval, G., Barreiro, L., Salas, A., Semino, O., Santachiara-Benerecetti, S., Kidd, K., Kidd, J., Van Der Veen, L., Hombert, J., et al. (2009). Inferring the demographic history of African farmers and pygmy hunter-gatherers using a multilocus resequencing data set. *PLoS Genetics*, 5(4):e1000448.

Pham, K. C., Nott, D. J., and Chaudhuri, S. (2014). A note on approximating ABC-MCMC using flexible classifiers. *Stat*, 3(1):218–227.

Prangle, D., Blum, M. G. B., Popovic, G., and Sisson, S. A. (2013). Diagnostic tools of approximate Bayesian computation using the coverage property. *ArXiv e-prints*.

Pritchard, J., Seielstad, M., Perez-Lezaun, A., and Feldman, M. (1999). Population growth of human Y chromosomes: a study of Y chromosome microsatellites. *Mol. Biol. Evol.*, 16:1791–1798.

Pudlo, P., Marin, J.-M., Estoup, A., Cornuet, J.-M., Gautier, M., and Robert, C. P. (2015). Reliable ABC model choice via random forests. *Bioinformatics*. (To appear.).

Ramakrishnan, U. and Hadly, E. (2009). Using phylochronology to reveal cryptic population histories: review and synthesis of 29 ancient DNA studies. *Molecular Ecology*, 18(7):1310–1330.

Ratmann, O., Andrieu, C., Wiujf, C., and Richardson, S. (2009). Model criticism based on likelihood-free inference, with an application to protein network evolution. *Proc. Natl. Acad. Sciences USA*, 106:1–6.

Robert, C. (2001). *The Bayesian Choice*. Springer-Verlag, New York, second edition.

Robert, C. (2012). Discussion of "constructing summary statistics for Approximate Bayesian Computation" by P. Fernhead and D. Prangle. *J. Royal Statist. Society Series B*, 74(3):447–448.

Robert, C. and Casella, G. (2004). *Monte Carlo Statistical Methods*. Springer-Verlag, New York, second edition.

Robert, C., Cornuet, J.-M., Marin, J.-M., and Pillai, N. (2011). Lack of confidence in ABC model choice. *Proceedings of the National Academy of Sciences*, 108(37):15112–15117.

Rubin, D. (1984). Bayesianly justifiable and relevant frequency calculations for the applied statistician. *Ann. Statist.*, 12:1151–1172.

Slatkin, M. (1995). A measure of population subdivision based on microsatellite allele frequencies. *Genetics*, 139(1):457–462.

Stephens, M. and Donnelly, P. (2000). Inference in molecular population genetics. *Journal of the Royal Statistical Society: Series B (Statistical Methodology)*, 62(4):605–635.

Stoehr, J., Pudlo, P., and Cucala, L. (2014). Adaptive ABC model choice and geometric summary statistics for hidden Gibbs random fields. *Statistics and Computing*, pages 1–13.

Sunnåker, M., Busetto, A., Numminen, E., Corander, J., Foll, M., and Dessimoz, C. (2013). Approximate Bayesian computation. *PLoS Comput. Biol.*, 9(1):e1002803.

Tanner, M. and Wong, W. (1987). The calculation of posterior distributions by data augmentation. *J. American Statist. Assoc.*, 82:528–550.

Tavaré, S., Balding, D., Griffith, R., and Donnelly, P. (1997). Inferring coalescence times from DNA sequence data. *Genetics*, 145:505–518.

Templeton, A. (2008). Statistical hypothesis testing in intraspecific phylogeography: nested clade phylogeographical analysis vs. approximate Bayesian computation. *Molecular Ecology*, 18(2):319–331.

Templeton, A. (2010). Coherent and incoherent inference in phylogeography and human evolution. *Proc. National Academy of Sciences*, 107(14):6376–6381.

Toni, T., Welch, D., Strelkowa, N., Ipsen, A., and Stumpf, M. (2009). Approximate Bayesian computation scheme for parameter inference and model selection in dynamical systems. *Journal of the Royal Society Interface*, 6(31):187–202.

Verdu, P., Austerlitz, F., Estoup, A., Vitalis, R., Georges, M., Théry, S., Froment, A., Le Bomin, S., Gessain, A., Hombert, J.-M., Van der Veen, L., Quintana-Murci, L., Bahuchet, S., and Heyer, E. (2009). Origins and genetic diversity of pygmy hunter-gatherers from Western Central Africa. *Current Biology*, 19(4):312–318.

Wegmann, D. and Excoffier, L. (2010). Bayesian inference of the demographic history of chimpanzees. *Molecular Biology and Evolution*, 27(6):1425–1435.

Wikipedia (2014). Approximate Bayesian computation — Wikipedia, The Free Encyclopedia.

Wilkinson, D. (2006). *Stochastic Modelling for Systems Biology*. Chapman & Hall/CRC Press, Boca Raton, Florida.

Wilkinson, R. (2013). Approximate Bayesian computation (ABC) gives exact results under the assumption of model error. *Statistical Applications in Genetics and Molecular Biology*, 12(2):129–141.

CLUSTERING MILKY WAY'S GLOBULAR CLUSTERS: A BAYESIAN NONPARAMETRIC APPROACH

Julyan Arbel [1]

Abstract. This chapter presents a Bayesian nonparametric approach
to clustering, which is particularly relevant when the number of components in the clustering is unknown. The approach is illustrated with
the Milky Way's globular clusters, that are clouds of stars orbiting in
our galaxy. Clustering these objects is key for better understanding
the Milky Way's history. We define the Dirichlet process and illustrate
some alternative definitions such as the Chinese restaurant process, the
Pólya Urn, the Ewens sampling formula, the stick-breaking representation through some simple R code. The Dirichlet process mixture model
is presented, as well as the R package BNPmix implementing Markov
chain Monte Carlo sampling. Inference for the clustering is done with
the variation of information loss function.

1 R requirements

The code used during the presentation of the Stat4Astro summer school is available
at the url: https://github.com/jarbel/Stat4Astro-Autrans. Additionally, the code
used to generate the plots of this chapter is displayed in the text. This requires
the following R packages: ggplot2, hexbin, viridis, gridExtra, ggpubr, rgl
for graphical tools, reshape2 for operations on similarity matrices, mclust and
mclust.ext for clustering estimation.

```
needed_packages <- c("ggplot2", "hexbin", "viridis", "gridExtra",
                     "ggpubr", "rgl", "reshape2", "dplyr", "mclust")
new_packages <- needed_packages[
  !(needed_packages %in% installed.packages()[, "Package"])]
if (length(new_packages))
  install.packages(new_packages)
lapply(needed_packages, require, character.only = TRUE)
```

[1] Univ. Grenoble Alpes, Inria, CNRS, Grenoble INP, LJK, 38000 Grenoble, France.

```
download.file(
  url = "http://wrap.warwick.ac.uk/71934/1/mcclust.ext_1.0.tar.gz",
  destfile = "mcclust.ext_1.0.tar.gz")
install.packages("mcclust.ext_1.0.tar.gz", repos = NULL, type = "source"
file.remove("mcclust.ext_1.0.tar.gz")
library("mcclust.ext")
```

2   Introduction and motivation

Globular clusters (hereafter globulars)[1] are sets of stars orbiting some galactic
center. The globular data we are considering here was studied in the 2015 Edition
of the Stat4Astro school by Fraix-Burnet et al. (2009) who used phylogenetic
classification. The data are available on GitHub and can be downloaded as follows:

```
spectra <- read.csv(
  "https://github.com/jarbel/Stat4Astro-Autrans/blob/
  master/Talk_Arbel/bnp_code/data/GC4c_groups.dat",
  sep="")
```

It lists a total of `dim(spectra)[1]`=54 globulars for which `dim(spectra)[2]`=7
variables are available[2]: GC stands for the globular identifier; logTe is the logarithm of the maximum effective temperature on the horizontal branch; FeH denotes
the metallicity; MV is the absolute V magnitude, which relates to both the bright-
ness and the mass of the globular; Age of the globular; Grp4c and Grp3c are the
phylogenetic classifications of Fraix-Burnet et al. (2009) obtained by using respectively the four variables logTe, FeH, MV, Age and the three variables logTe,
FeH, MV.

```
## [1] 54  7
```

```
## [1] "GC"     "logTe" "FeH"    "MV"     "Age"    "Grp4c" "Grp3c"
names(spectra)
```

```
## [1] "GC"     "logTe" "FeH"    "MV"     "Age"    "Grp4c" "Grp3c"
```

By using the additional spatial coordinates available on the Wikipedia list[3], we
can obtain a spatial scatterplot of the globulars. In Figure 1, the globulars present
in the study (that is for which we have measurements for the above mentionned
variables) are depicted in purple, the others in green

The two clusterings Grp4c and Grp3c are of respective size 3 and 4, and the
cluster sizes are obtained with the table command as follows:

---

[1]Globulars are more commonly called globular clusters in the literature, though we shall prefer
the phrasing 'globulars' to 'globular clusters' in order to avoid ambiguous terms like 'globular

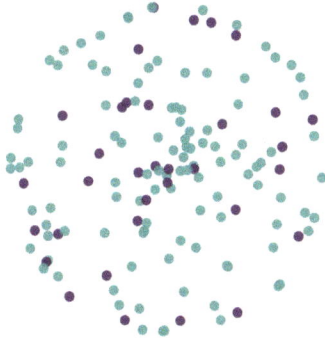

Fig. 1. Globular coordinates. Gobulars included in the present study are depicted in purple, the others in green

```
table(spectra$Grp3c)
```

```
##
##  1  2  3  4
## 17  7 21  9
```

```
table(spectra$Grp4c)
```

```
##
##  1  2  3
## 25 11 18
```

From now on, we will focus on the three variables `logTe, FeH, MV` only. A 2-D representation for each pair of variables, as well as a side density plot for each variable can be obtained by:

```
col <- plasma(1)
alpha <- .8
spVF <- ggscatter(spectra, x = "MV", y = "FeH", color = col,
                  size = 3, alpha = alpha)+ border()
spLF <- ggscatter(spectra, x = "logTe", y = "FeH", color = col,
                  size = 3, alpha = alpha)+ border()
spLV <- ggscatter(spectra, x = "logTe", y = "MV", color = col,
```

---

clusters clusters'.

[2]A more comprehensive list of globulars can be accessed at https://en.wikipedia.org/wiki/List_of_globular_clusters, though it contains information about magnitude and diameter only.

[3][2]

```
                        size = 3, alpha = alpha) + border()
Vplot <- ggdensity(spectra, "MV", fill = col)
Lplot <- ggdensity(spectra, "logTe", fill = col)
Fplot <- ggdensity(spectra, "FeH", fill = col) + rotate()
Vplot <- Vplot + clean_theme()
Lplot <- Lplot + clean_theme()
Fplot <- Fplot + clean_theme()
ggarrange(Vplot, Lplot, NULL,
          spVF, spLF, Fplot,
          NULL, spLV, NULL,
          ncol = 3, nrow = 3,  align = "hv",
          widths = c(2, 2, 1), heights = c(1, 2, 2))
```

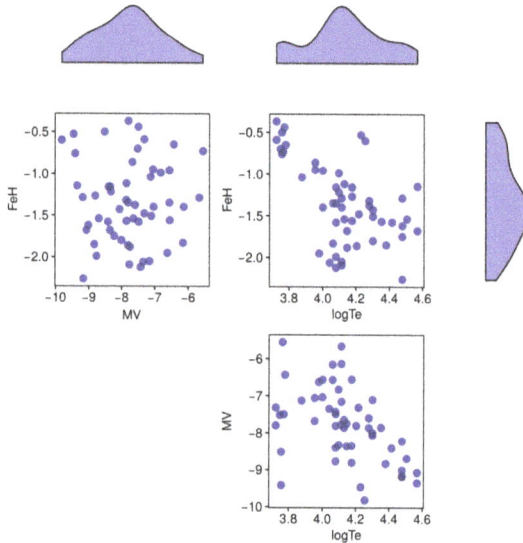

These scatter plots hardly identify any clustering structure amongst the globulars. However, astrophysicists expect several populations of globulars identified by similar time of formation, and similar chemical and physical conditions. The granularity of the populations of globulars may depend on the size of the considered sample: a small number of observations would likely lead to little discriminating power, while a large sample size would provide more evidence for identifying more distinct clusters. This setting where the number of clusters might grow with the sample size is well suited to a Bayesian nonparametric approach to clustering, which this chapter aims at introducing.

The rest of the chapter is organised as follows. Model-based clustering and the Dirichlet process are introduced in Section 3 and Section 4. We conclude with an illustration to the Globular dataset.

3 Model-based clustering

Mixture models are creating flexible models starting from simple ones. For instance, combining two unimodal densities p_1 and p_2 into $\pi p_1 + (1 - \pi)p_2$, for $\pi \in (0, 1)$ can create a bimodal distribution. The aim of mixtures is to increase the modelling capacities by combining simple distributions into flexible distributions which might display multimodality, skewness, etc. A mathematical definition of a mixture density is a convex combination of densities. Each density can be interpreted as a sub-population. Observations are associated to sub-populations through latent (un-observed) variables called allocation variables, which play a major role in devising sampling algorithms.

Consider a parametric family of distributions with densities $p(\cdot|\phi)$, ϕ element of some parameter space Θ, with respect to a common measure. Broadly speaking, mixtures operate in a discrete way (finite or infinite countable number of classes) or in a continuous way (infinite uncountable number of classes). As an example of continuous mixtures, a Student t-distribution can be written as a mixture of Gaussian kernels with fixed mean and variance averaged over an inverse gamma distribution (playing the role of a mixing distribution). We will instead focus here on discrete mixtures.

Going back to the simple mixture of two densities, the mixture density takes on the form $\pi p_1(\cdot|\phi_1) + (1 - \pi)p_2(\cdot|\phi_2)$, where the unknown parameters are (π, ϕ_1, ϕ_2). Modelling data $\mathbf{x} = (x_1, \ldots, x_N)$ with such a density can be thought as the following two-step procedure, for $i \in \{1, \ldots, N\}$:

1. toss a coin, which samples from one of the two classes $\theta_i \in \{\phi_1, \phi_2\}$. In other words, the class identified with parameter ϕ_1 has probability π, while the ϕ_2 class has the complement probability $1 - \pi$;

2. sample the observation from the corresponding density $x_i \sim p(x|\theta_i)$.

The coin-tossing step can be equally thought of as a throw of dice whose number of facets equals the number of components in the mixture, and each facet has a probability given by the component weight. This mixing distribution takes the mathematical form of a convex combination $\pi \delta_{\phi_1} + (1 - \pi)\delta_{\phi_2}$ of Dirac masses at ϕ_1 and ϕ_2, each of which identifies with a class of the mixture.

We speak about nonparametric (discrete) mixtures when the number of classes is (countable) infinite. In which case the dice has an infinite number of facets. We index the classes by the positive integers, and denote the parameters by ϕ_k and weights by π_k. Then those weights must sum up to one, and the mixing measure can be written as

$$G = \sum_{k=1}^{\infty} \pi_k \delta_{\phi_k},$$

which is a probability measure. Sampling from the mixture distribution

$$p_G(\cdot) = \sum_{k=1}^{\infty} \pi_k p(\cdot|\phi_k)$$

can again be done in the sequential way, for $i \in \{1, \ldots, N\}$:

1. throw the infinite dice G: $\theta_i \sim G$;

2. sample the observation from the corresponding density $x_i \sim p(x|\theta_i)$.

The mixture density p_G has an infinite number of parameters: the vector of weights $(\pi_1, \pi_2, \ldots)$ which is an element of the infinite simplex, and the vector of class parameters $(\phi_1, \phi_2, \ldots)$. A Bayesian approach to the problem requires endowing the parameters with some prior distribution. The Dirichlet process is the most prominent example of such a prior distribution. It is parametrised by the base measure, denoted by G_0, and the precision parameter, a positive scalar denoted by α. Figure 2 displays plate representations of a Bayesian and a non Bayesian (or frequentist) approach to the mixture model. Such a plate representation makes it clear that a Bayesian approach add a hierarchical layer (that of the prior) to the non Bayesian set-up.

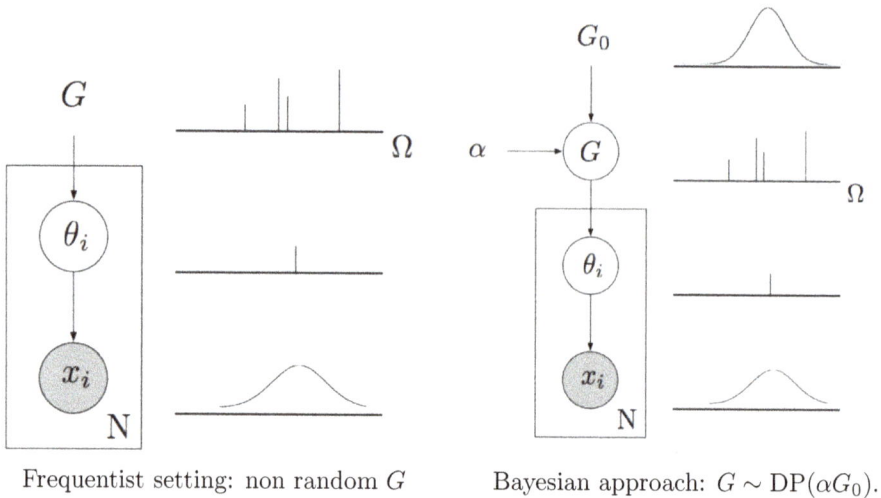

Frequentist setting: non random G Bayesian approach: $G \sim \mathrm{DP}(\alpha G_0)$.

Fig. 2. Plate representations of a nonparametric mixture with mixing measure G.

Choosing the number of components to include in a finite mixture is often a delicate question, as one rarely knows it a priori. There are mainly three strategies: i) to fit several finite mixture models with a range of plausible values for the number of components; then choose the model which maximises some information criterion; ii) consider the number of components as Bayesian parameter, which is to say endow it with a prior distribution on positive integers, such as a Poisson distribution. Such a model is termed a mixture of finite mixtures (MFM) by Miller and Harrison (2017); or iii) let this number be a priori infinite (or as large as needed) and select it or estimate it a posteriori. We focus on this last option in the next section.

4   Bayesian nonparametrics around the Dirichlet process

Most of the models described in the previous chapters of this book are parametric: they can be described by a finite and fixed number of parameters. This number of parameters is independent of the dataset. Very convenient models in a Bayesian setting include conjugate models, where the posterior distribution has the same form as the prior distribution. For example, normal prior and normal likelihood, or beta prior and binomial likelihood, or gamma prior and Poisson likelihood. In contrast, we are dealing in this chapter with *nonparametric* models. Let us put it straightaway, the *nonparametric* saying is not the most fortunate, as nonparametric models *do* have parameters, many of them. Bernardo and Smith (2009) define a *Bayesian nonparametric model* as a probability model with infinitely many parameters, also referred to by Müller and Mitra (2013) as a model with massively many parameters. I think there are three ways this large number of parameters can be thought of: the number of parameters is *infinite*, or it is *random*, or it is *growing* with the data sample size.

It goes without saying that parametric models are easier to handle than their nonparametric counterparts, that are computationally and analytically more challenging. It is also true that interpreting a small and fixed number of parameters is likely to be easier than for, say, an infinite number of them. However, the computational and analytical extra burden is the price to pay for flexibility. Nonparametric models are less prone to misspecification than parametric models, which require a strong belief in the particular structure they imply. Without such a belief in the parametric assumption, a model might not be reliable. By nature, nonparametric models are well suited to study curves in general, so typical applications include density estimation and regression estimation. Clustering is also an example which has historically played a central role within Bayesian nonparametrics.

In this section we provide basic facts about the Dirichlet process, precisely different possible representations: the definition via the finite dimensional marginal distribution in Definition 4.2, the Chinese restaurant process in Proposition 4.3, the posterior distribution in Theorem 4.1, the Pólya Urn in Proposition 4.2 and the stick-breaking representation in Theorem 4.2.

4.1   Definition

The Dirichlet process (Ferguson, 1973) plays a central role in Bayesian nonparametrics. A Dirichlet process can be viewed as a random variable where the variable is a probability measure. It has two parameters: the base measure, denoted by G_0, and the precision parameter, a positive scalar denoted by α.

(Ferguson, 1973) defines the Dirichlet process by its finite marginal distributions: i.e., how does the random measure spread its mass in the sample space? The answer is: as a Dirichlet distribution (which, in passing, explains the name). Recall that the Dirichlet distribution generalises the Beta distribution to any dimension $k \geq 2$. The Dirichlet distribution in $\mathbb{R}^k$ is restricted to the unit simplex, with density parametrised by k positive scalars $\alpha_1, \ldots, \alpha_k$ and proportional to

$$x_1^{\alpha_1-1} \cdots x_k^{\alpha_k-1}.$$

Definition 4.1 (Dirichlet distribution). *A Dirichlet distribution on a simplex Δ_K is a probability distribution with parameters $\alpha_i > 0$ and a density function*

$$f(x_1, \ldots, x_K; \alpha_1, \ldots, \alpha_K) = \frac{1}{B(\alpha)} \prod_{i=1}^{K} x_i^{\alpha_i-1}.$$

It is common to refer to Dirichlet distribution as $\mathrm{Dir}(x_1, \ldots, x_k)$. Let us remark that

Remark 4.1. *The Dirichlet distribution is conjugate prior for the multinomial distribution.*

Consider a finite partition of the space, denoted by $(A_1, \ldots, A_k)$. The mass allocated by a Dirichlet process G to each region A_j is a random variable $G(A_j)$, thus giving rise to a random vector $G(A_1), \ldots, G(A_k)$ for the whole partition. Then G follows a Dirichlet process with parameters α and G_0 if the random vector $G(A_1), \ldots, G(A_k)$ is distributed as a Dirichlet distribution with parameters $\alpha G_0(A_1), \ldots, \alpha G_0(A_k)$.

Definition 4.2 (Dirichlet process, Ferguson (1973)). *A random probability measure G follows a Dirichlet process with parameters α and G_0 on some space if for any finite partition $(A_1, \ldots, A_k)$ of the space,*

$$(G(A_1), \ldots, G(A_k)) \sim \mathrm{Dir}(\alpha G_0(A_1), \ldots, \alpha G_0(A_k)).$$

Note that this definition entails the strong result that such finite dimensional distributions consistently define a stochastic process. Figure 3 shows different realisations of a Dirichlet process in $\mathbb{R}^2$, with a standard Gaussian base measure G_0 and precision parameter equal to one, and with different partitions. Of course, not all of the sample space $\mathbb{R}^2$ can be represented, but most of the mass is captured in the represented part of the space.

4.2 Properties

From Definition 4.2, we see that any measure set A of the space receives mass according to the following Beta distribution

$$P(A) \sim \mathrm{Beta}(\alpha G_0(A), \alpha(1 - G_0(A)). \tag{4.1}$$

We have the following moments.

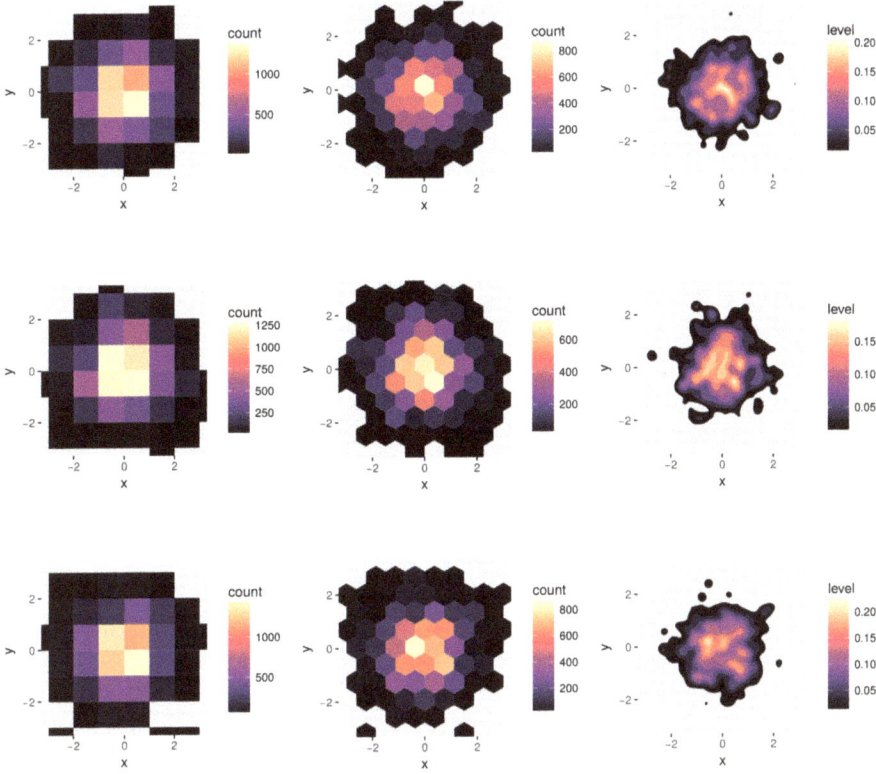

Fig. 3. Dirichlet process realisations in $\mathbb{R}^2$, varying in rows. Different partitions of $\mathbb{R}^2$ are used in columns. The base measure G_0 is a standard Gaussian and the precision parameter is equal to one.

Proposition 4.1 (Dirichlet process moments). *If $G \sim \mathrm{DP}(\alpha G_0)$, then for any measurable sets A and B*

$$\mathrm{E}[G(A)] = G_0(A)$$

$$\mathrm{Var}[G(A)] = \frac{G_0(A)(1 - G_0(A))}{1 + \alpha}$$

$$\mathrm{Cov}[G(A), G(B)] = \frac{G_0(A \cap B) - G_0(A)G_0(B)}{1 + \alpha}$$

Proof. We will make use of Equation (4.1). From this we obtain

$$\mathrm{E}(P(A)) = \frac{\alpha P_0(A)}{\alpha(P_0(A) + 1 - P_0(A))} = P_0(A)$$

and

$$\mathrm{Var}(P(A)) = \frac{\alpha^2 P_0(A)(1 - P_0(A))}{\alpha^2(\alpha + 1)}.$$

We derive the covariance term in two cases, firstly taking into consideration the one with $A \cap B = \emptyset$. In that case the sample space is partitioned into A, B and $(A \cup B)^c$, the complementary set of $A \cup B$, which is equal to $(A \cup B)^c = A^c \cap B^c$. Therefore we may write a joint probability vector

$$\big(P(A), P(B), P(A^c \cap B^c)\big) \sim \mathrm{Dir}\big(\alpha P_0(A), \alpha P_0(B), \alpha P_0(A^c \cap B^c)\big),$$

and hence $\mathrm{Cov}(P(A), P(B)) = -P_0(A)P_0(B)/(1 + \alpha)$. In the more general case one may decompose

$$A = (A \cap B) \cup (A \cap B^c)$$
$$B = (B \cap A) \cup (B \cap A^c),$$

so that

$$\mathrm{Cov}(P(A), P(B)) = \mathrm{Cov}(P(A \cap B) + P(A \cap B^c), P(B \cap A) + P(B \cap A^c))$$

and so forth using the linearity of covariance.   □

In passing, this shows that any two measurable parts of the space which are non intersecting receive mass from the Dirichlet process with negative correlation. Which makes sense since the total measure is constrained to be a probability measure, so more mass in some part of the space means less in the rest of the space.

Another central property is the conjugacy of the Dirichlet process: if some data are sampled from a Dirichlet process G, then the posterior distribution of G conditional on the data is still a Dirichlet process.

Theorem 4.1 (Dirichlet process posterior distribution, Ferguson (1973)). *Let data* $\boldsymbol{X} = (X_1, \ldots, X_n)$ *be distributed according to the model*

$$P \sim \mathrm{DP}(\alpha G_0)$$
$$X_1, \ldots, X_n | G \overset{iid}{\sim} G.$$

Then the posterior distribution of G is given by

$$G | X_1, \ldots, X_n \sim \mathrm{DP}\left(\alpha P_0 + \sum_{i=1}^{n} \delta_i\right).$$

Proof. This posterior can be obtained by remarking that for any finite measurable partition $(A_1, \ldots, A_k)$, the posterior distribution of $P(A_1), \ldots, P(A_k)$ depends on the observations only via their cell counts (this comes from the *tail-free* property of the DP). Denote $N_j = \#\{1 \leq i \leq n : x_i \in A_j\}$, i.e. the number of observations in each cell of the partition of $(A_1, \ldots, A_k)$. Then we have

$$\big(P(A_1), \ldots, P(A_k)\big)\big|X_{1:n} \overset{\mathrm{d}}{=} \big(P(A_1), \ldots, P(A_k)\big)\big|N_{1:k}.$$

Let us use the shorthand notation: $\boldsymbol{\alpha} = (\alpha_1, \ldots, \alpha_k) = (P(A_1), \ldots, P(A_k))$ and $\boldsymbol{N} = (N_1, \ldots, N_k)$. Then

$$\begin{cases} \boldsymbol{\alpha} \sim \mathrm{Dir}_k(\alpha P_0(A_1), \ldots, \alpha P_0(A_k)), \\ \boldsymbol{N}|P \sim \mathrm{Multinom}_k(n, \boldsymbol{\alpha}), \end{cases}$$

and hence we obtain the prior of the form

$$p(\boldsymbol{\alpha}) \propto \alpha_1^{\alpha P_0(A_1)-1} \cdots \alpha_k^{\alpha P_0(A_k)-1},$$

while sampling model is

$$p(\boldsymbol{N}|\boldsymbol{\alpha}) \propto \alpha_1^{N_1} \cdots \alpha_k^{N_k}.$$

This results in the posterior of form

$$p(\boldsymbol{\alpha}|\boldsymbol{N}) \propto \alpha_1^{\alpha P_0(A_1)+N_1-1} \cdots \alpha_k^{\alpha P_0(A_k)+N_k-1} = \mathrm{Dir}_k\big(\alpha P_0(A_1)+N_1, \ldots, \alpha P_0(A_k)+N_k\big).$$

$\square$

Note that the notational convention adopted for the parameters of the Dirichlet process is interesting here: it provides the simple 'product' parameter $\alpha P_0 + \sum_{i=1}^{n} \delta_i$, which can also be decoupled into a posterior precision parameter α_n and a posterior base measure G_n given by

$$\alpha_n = \alpha + n,$$

$$G_n = \frac{\alpha}{\alpha + n} P_0 + \frac{1}{\alpha + n} \sum_{i=1}^{n} \delta_{X_i}.$$

This posterior conjugacy, in turn, provides a simple form of the predictive distribution, that is the distribution of a new observation conditional on the data. This predictive is referred to as the Pólya urn, or Blackwell-MacQueen urn.

Proposition 4.2 (Predictive distribution, Pólya urn, Blackwell and MacQueen (1973)). *In the model of Proposition 4.1, the predictive distribution for a new observation X_{n+1} is given by*

$$X_{n+1}|X_1, \ldots, X_n \sim \frac{\alpha}{\alpha + n} P_0 + \frac{1}{\alpha + n} \sum_{i=1}^{n} \delta_i.$$

Proof. This property is a result of taking the expected value of the posterior given in Proposition 4.1. $\square$

Such a predictive distribution induces ties in the observations with positive probability. More precisely, n observations sampled from a DP induce a partition of the integers $1, \ldots, n$. The distribution of this random partition is called the Chinese restaurant process[4]. This culinary metaphor describes the random partition induced by the DP as follows. Customers join a populated table with probability $n_j/(\alpha + n)$, where n_j denotes the number of clients already sitting around the table or sit at new table with probability $\alpha/(\alpha + n)$.

Proposition 4.3 (Chinese restaurant process, Antoniak (1974)). *A random sample $X_{1:n}$ from a DP with precision parameter α induces a partition of $\{1, \ldots, n\}$ into k sets of sizes $n_1, \ldots, n_k$ with probability*

$$p(n_1, \ldots, n_k) = p(\{n_1, \ldots, n_k\}) = \alpha^k \frac{\Gamma(\alpha)}{\Gamma(\alpha + n)} \prod_{j=1}^{k} \Gamma(n_j).$$

Proof. We will use the Pólya urn schema slightly changed by using $n_1, \ldots, n_k$

$$P(X_{n+1}|X_{1:n}) = \frac{\alpha}{\alpha + n} P_0 + \frac{1}{\alpha + n} \sum_{j=1}^{k} n_j \delta_{X_j^*}.$$

By exchangeability, the distribution of $\{n_1, \ldots, n_k\}$ does not depend on the order of the observations. Let's compute $p(n_1, \ldots, p_k)$ as the probability of one draw where the first table consists of first n_1 observations etc.

To proceed, let us use Pólya urn scheme: we denote $\bar{n}_j = \sum_{i=1}^{j} n_i$ and hence $\bar{n}_k = n$, the total number of observations. We can observe the following pattern: first ball opens new table, following $n_j - 1$ ones fill in that table and so forth. That quantity can be rewritten as

$$\frac{\alpha^k}{\alpha(\alpha + 1) \ldots (\alpha + n - 1)} \prod_{j=1}^{k} (n_j - 1)!,$$

where one can rewrite both terms using Gamma function $\Gamma(x) = \int_0^\infty u^{x-1} e^{-u} du$:

$$\alpha(\alpha + 1) \ldots (\alpha + n - 1) = \frac{\Gamma(\alpha + n)}{\Gamma(\alpha)},$$

and $(n_j - 1)! = \Gamma(n_j)$.

Note that for ordered partitions we have

$$\bar{p}(n_1, \ldots, n_k) = \frac{p(n_1, \ldots, n_k)}{k!}.$$

□

---

[4]According to Aldous (1985), the restaurant analogy is due to Jim Pitman and Lester Dubins.

The following lines of code sample observations from a Dirichlet process with a base measure in argument, which is also interpreted as the color distribution in the Pólya urn scheme.

```
polya_urn_model <- function(base_measure, N_ball, alpha) {
  balls <- c()
  for (i in 1:N_ball) {
    if (runif(1) < alpha / (alpha + length(balls))) {
      # Add a new ball color.
      new_color <- base_measure()
      balls <- c(balls, new_color)
    } else {
      # Pick out a ball from the urn, and add back a
      # ball of the same color.
      ball <- balls[sample(1:length(balls), 1)]
      balls <- c(balls, ball)
    }
  }
  balls
}
```

This is applied to sample 10 observations from a Dirichlet process with a Gaussian $N(0, 1)$ base measure, which is also interpreted as the color distribution in the Pólya urn scheme, and precision parameter varying from 1, 10 to 100.

```
N_ball <- 10
# with alpha = 1
polya_sample <- polya_urn_model(function() rnorm(1), N_ball, 1)
rev(sort(table(polya_sample)))

## polya_sample
## -1.04779556205753 0.353418450483626
##                 8                  2

# with alpha = 10
polya_sample <- polya_urn_model(function() rnorm(1), N_ball, 10)
rev(sort(table(polya_sample)))

## polya_sample
##  -0.262911838454736   -0.85109415351288    2.01291826160114
##                   2                   2                   1
##    1.73642130785405   0.546212584065414   0.264154437076562
##                   1                   1                   1
##   0.191324198334465 0.00291702961916057
##                   1                   1
```

```
# with alpha = 100
polya_sample <- polya_urn_model(function() rnorm(1), N_ball, 100)
rev(sort(table(polya_sample)))
```

```
## polya_sample
##   0.977248376259983   0.766722814230755   0.597919221042819
##                   1                   1                   1
##   0.466119538300814   0.460876928395281   0.304868236499072
##                   1                   1                   1
##   0.157307784983287  -0.269715738859888  -0.652901580743433
##                   1                   1                   1
##  -1.41731999466404
##                   1
```

These experiments illustrate that large values of the mass parameter α tend to produce a larger number of distinct values in a sample of a given size. For instance, the probability that all $n = 10$ observations are distinct is equal to

$$\mathbb{P}(X_1, \ldots, X_n \text{ are pairwise distinct}) = \alpha^n \frac{\Gamma(\alpha)}{\Gamma(\alpha + n)},$$

which is approaching 1 as α grows, for n fixed.

In our case, this probability is respectively equal to 3e-07, 3e-02 and 0.6 for α equal to 1, 10 and 100, for $n = 10$.

4.3 Stick-breaking representation

The Dirichlet process is a discrete random probability measure which can be represented as a convex combination of infinitely many Dirac masses,

$$G = \sum_{k=1}^{\infty} \pi_k \delta_{\phi_k}.$$

The stick-breaking representation, due to Sethuraman (1994), provides a constructive way of building the weights $(\pi_k)_k$ of the Dirichlet process. This is done by sequentially breaking a stick of initial unit length, into pieces whose lengths correspond to the $(\pi_k)_k$. More specifically, we require independent and identically distributed (iid) random variables $V_k \sim \text{Beta}(1, \alpha)$. The first weight π_1 corresponds to V_1. This leaves a piece of length $1 - V_1$, which is broken at V_2 in order to define $\pi_2 = V_2(1 - V_1)$. And sequentially, the same procedure is applied to the remaining part, which equals $(1 - V_1)(1 - V_2)$ at this second step. It is easy to see that after k steps, one defines $\pi_k = V_k(1 - V_1) \cdots (1 - V_{k-1})$, and the remaining piece has length $(1 - V_1) \cdots (1 - V_k)$. The representation is completed by assuming iid draws from the base measure G_0 for the locations ϕ_k, independent from the V_k.

Theorem 4.2 (Stick-breaking representation, Sethuraman (1994)). *Let $V_1, V_2, \dots \overset{iid}{\sim}$ Beta$(1, \alpha)$ and $\phi_1, \phi_2, \dots \overset{iid}{\sim} G_0$ be independent random variables. Define*

$$\pi_1 = V_1,$$
$$\pi_k = V_k(1 - V_1) \cdots (1 - V_{k-1}), \text{ for any } k \geq 2.$$

Then $G = \sum_{k=1}^{\infty} \pi_k \delta_{\phi_k} \sim \mathrm{DP}(\alpha G_0)$.

Proof. We provide a sketch of proof of this result in two steps. First, to show that the remaining stick length at step k, $(1 - V_1) \cdots (1 - V_k)$, converges to zero as $k \to \infty$. This ensures that the weights vector lives in the unit simplex, and in turn that the measure $\sum_{k=1}^{\infty} \pi_k \delta_{\phi_k}$ is a probability measure. Second, to use the stick-breaking construction to show that the defined G satisfies the distributional equation

$$G \overset{d}{=} V\delta_\phi + (1 - V)G, \tag{4.2}$$

where $V \sim$ Beta$(1, \alpha)$ and $\phi \sim G$, independently, whose only solution turns out to be the Dirichlet process, by properties of the Dirichlet distribution. □

The following code implements the sampling of the first `num_weights=50` weights of a DP. These are then plotted in the order they appear in the stick-breaking construction (left) as well as indexed by their corresponding location (right). Colors indicate the stick-breaking order of appearance. Note that the weights are not necessarily strictly decreasing, but only stochastically decreasing. This means they are decreasing in expectation, as can be easily checked:

$$\mathrm{E}(\pi_k) = \mathrm{E}(V_k(1-V_1) \cdots (1-V_{k-1})) = \mathrm{E}V_k\mathrm{E}(1-V_1) \cdots \mathrm{E}(1-V_{k-1}) = \frac{1}{\alpha + 1}\left(\frac{\alpha}{\alpha + 1}\right)^{k-1}.$$

```
stick_breaking_process = function(num_weights, alpha) {
  betas = rbeta(num_weights, 1, alpha)
  remaining_stick_lengths = c(1, cumprod(1 - betas))[1:num_weights]
  weights = remaining_stick_lengths * betas
  weights
}
```

```
num_weights <- 50
draw_stick_breaking <- function(alpha) {
  labels <- 1:num_weights
  locations <- rnorm(num_weights)
  SB_weights <- stick_breaking_process(num_weights, alpha)
  df <- data.frame(labels, locations, SB_weights)
  order_plot <-
        ggplot(df, aes(labels, SB_weights, fill = as.factor(labels))) +
```

```
    geom_bar(stat = "identity") +
    theme(legend.position="none")
  location_plot <-
      ggplot(df, aes(locations, SB_weights, fill = as.factor(labels)))
    geom_bar(stat = "identity", width = .1) +
    theme(legend.position="none")
  grid.arrange(order_plot, location_plot, ncol = 2)
}
```

draw_stick_breaking(1)

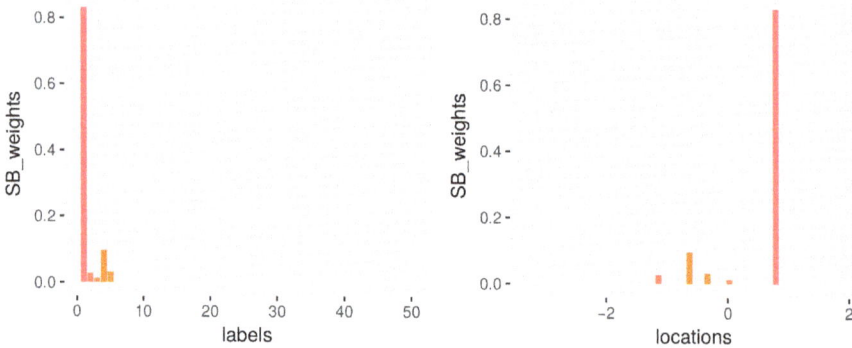

draw_stick_breaking(10)

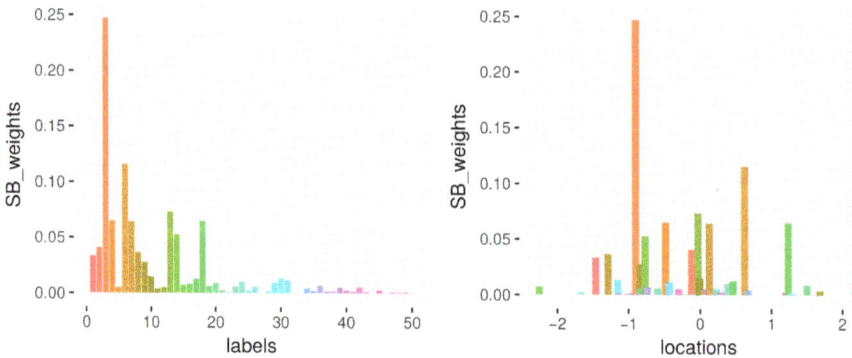

draw_stick_breaking(100)

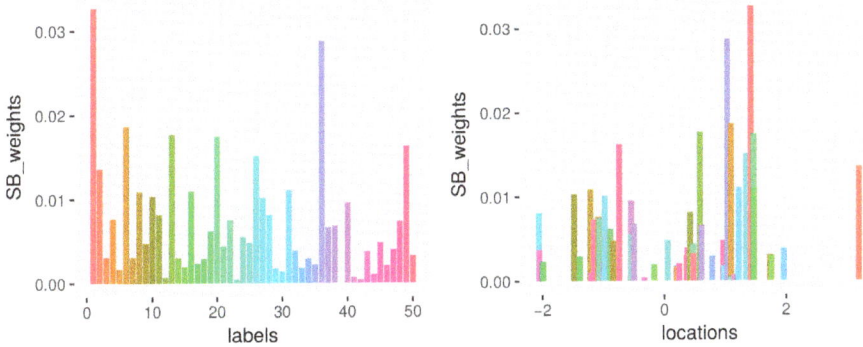

4.4   Dirichlet process mixture models

Combining the model-based clustering approach described in Section 3 with the DP, we are ready to work with nonparametric mixture models, where the Dirichlet process can be used as a prior distribution on the mixing probability measure (Lo, 1984).

The following code shows densities draws from DP mixture densities with $N(0,1)$ base measure, varying mass parameter in $\{1, 10, 100, 1000\}$, and centered Gaussian kernel with varying variance σ^2 in $\{0.2, 0.4, 1\}$. Enlarging α tends to produce flatter densities, while reducing σ^2 tends to produce more irregular ones.

```r
alpha_vect <- c(1, 10, 100, 1000)
N_urns <- 3
sigma2 <- c(1,.4,.2)
N_draws <- 100
N_xaxis <- 200
x_axis <- seq(-3, 3, length = N_xaxis)
result <- NULL
  for (alpha in alpha_vect) {
    PU <- polya_urn_model(function() rnorm(1), N_draws, alpha)
    for (u in 1:N_urns) {
    res <- mapply(function(mean) dnorm(x_axis, rep(mean, N_xaxis),
                                  rep(sigma2[u], N_xaxis)), PU)
    res <- apply(res, 1, mean)
    new_draw <- cbind(res, x_axis, alpha, sigma2[u])
    result <- rbind(result, new_draw)
  }
}
result <- as.data.frame(result)
names(result) <- c("density", "x", "alpha", "sigma2")
DP_mixt <- qplot(data = result, y = density, x = x,
                geom = c("line", "area")) +
```

```
facet_grid(alpha ~ sigma2, labeller =
              label_bquote(rows = alpha == .(alpha),
                            cols = sigma^2 == .(sigma2))) +
  aes(color = as.factor(alpha)) +
  theme(legend.position = "none")
DP_mixt
```

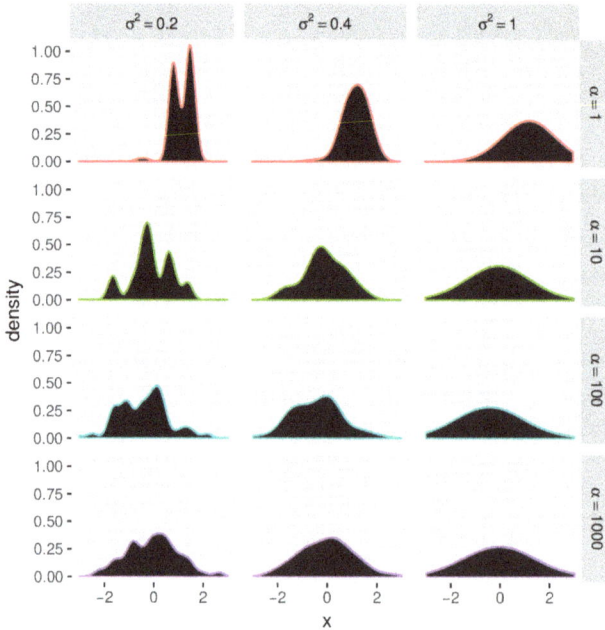

5   Application to clustering of globulars of our galaxy

In this last section, we show how Dirichlet process mixtures can be applied to perform clustering of globulars of our galaxy.

5.1   Markov chain Monte Carlo sampling

Inference in R for DPM is implemented in packages including **DPpackage** (Jara et al., 2011), **BNPdensity** (Barrios et al., 2013). Here we present a recent package called **BNPmix** (Arbel et al., 2018b). It uses C++ code with the **Rcpp** package, which makes the implementation efficient. The package is available on GitHub[5] and can be installed with **devtools** as follows:

---

[5]At the url: https://github.com/rcorradin/BNPmix.

```
library(devtools)
install_github("rcorradin/BNPmix")
```

This package uses d-dimensional location-scale Dirichlet process mixture of Gaussians. This means that it assumes a Gaussian kernel $p(X|\phi) = \Phi_d(X|\boldsymbol{\mu}, \boldsymbol{\Sigma})$, where the component specific parameter θ is composed of the mean $\boldsymbol{\mu}$ and covariance matrix $\boldsymbol{\Sigma}$. The base measure G_0 is chosen as an independent product of a Gaussian and an inverse-Wishart distributions

$$G_0(d\boldsymbol{\mu}, d\boldsymbol{\Sigma}; \boldsymbol{\pi}) = N_d(d\boldsymbol{\mu}; \mathbf{m}_0, \mathbf{B}_0) \times IW(d\boldsymbol{\Sigma}; \nu_0, \mathbf{S}_0), \qquad (5.1)$$

The hyperparameters $(\mathbf{m}_0, \mathbf{B}_0, \nu_0, \mathbf{S}_0)$ are typically expressed by an empirical Bayes approach. See (Arbel et al., 2018b) for details. Additionally, the precision parameter α is endowed with a gamma distribution with shape t_1 and scale t_2.

Posterior inference is carried out by a Markov chain Monte Carlo (MCMC) algorithm. The function is called `DPmixMulti`, and it takes the following arguments:

- the `data`;

- a `grid` of the sample space, with the same dimension as that of the data, on which the density is to be evaluated;

- `MCMC_param`, a list of parameters for the MCMC including number of simulations `nsim` and burn-in `nburn`;

- `starting_val`, a list containing the initial values for the components;

- `params`, hyperparameters $(\mathbf{m}_0, \mathbf{B}_0, \nu_0, \mathbf{S}_0)$ for the base measure G_0 of the DPM and (t_1, t_2) for the hyperprior on the precision parameter α. When omitted, those parameters are set in an empirical Bayes way by default.

A grid to evaluate the densities is defined as follows, spanning all data points, plus/minus 10% with respect to the extreme data points.

```
data <- spectra[, c(2,3,4)]
grid <- expand.grid(seq(range(data[,1])[1]-.1 * diff(range(data[,1])),
                        range(data[,1])[2]+.1 * diff(range(data[,1])),
                        length.out = 40),
                    seq(range(data[,2])[1]-.1 * diff(range(data[,2])),
                        range(data[,2])[2]+.1 * diff(range(data[,2])),
                        length.out = 40),
                    seq(range(data[,3])[1]-.1 * diff(range(data[,3])),
                        range(data[,3])[2]+.1 * diff(range(data[,3])),
                        length.out = 40))
```

We can now run the MCMC function `DPmixMulti`:

```
MCMC_param <-  list(nsim = 10^4, nburn = 5*10^3)
MCMC_output <- DPmixMulti(data = as.matrix(data),
                          grid = grid,
                          MCMC_param = MCMC_param)
```

The output `MCMC_output` of this function is a list of three objects:

- `distribution` the estimated (posterior mean) distribution evaluated on the grid.

- `result_cluster` a matrix, each row is an iteration, each column an observation, each entry is a latent component, thus this contains clusterings for each iteration of the chain.

- `result_theta` a vector, the value of DP precision parameter α over the iterations.

The posterior clustering information is enclosed in the second object listed above, that we can save in a matrix `MCMC_clustering`:

```
MCMC_clustering <- MCMC_output[[2]]
```

5.2 Clustering estimation

It should be noted that a DPM assumes a priori an infinite number of components in the mixture (see for instance the stick-breaking representation). As such, it is a misspecified model in essence when it comes to estimating a fixed clustering, that is a clustering which is not deemed to grow as the sampled data grows. However, in the spirit of the famous quote by George Box *"All models are wrong but some are useful"*, misspecified models can be used as long as they provide insightful results.

Clustering with DPM can be done in different ways. Indeed, the MCMC output we have obtained so far essentially consists in many different clusterings of the data, among which one needs to choose a 'best' clustering for some decision rule.

A first approach consists in looking at the number of components, an estimator of which can be obtained with the mode a posterior (MAP) for instance. However, the posterior distribution of the number of components in a DPM turns out to be *inconsistent* under some model specifications (Miller and Harrison, 2013). Posterior consistency is a theoretical property of a posterior distribution: when more and more data are collected from some fixed data distribution, we say that the posterior distribution of some parameter is consistent if it converges to a point mass at the true fixed value of this parameter. The inconsistency property provided by Miller and Harrison (2013) is as follows. Assume a DPM model with standard Gaussian base measure, precision parameter equal to one, and standard Gaussian kernel. Suppose data are generated from a standard Gaussian. This means that the data are sampled from a very simple mixture which admits a single component. Miller and Harrison (2013) study the posterior distribution of the number of clusters in this situation, and they prove that it does not concentrate to a point mass at one, thus proving posterior inconsistency.

Let us plot the histogram of the number of clusters:

```
MCMC_number_cluster = apply(X = MCMC_clustering, MARGIN = 1, FUN = max)
df_cluster <- data.frame(nb = MCMC_number_cluster)
c <- ggplot(df_cluster, aes(nb,..density..))
c + geom_histogram(breaks=(0:max(df_cluster)),
                   aes(fill=..density..)

                   ,
                   col="white",
                   alpha = .9
                   ) +
    scale_fill_viridis(option = "inferno") +
    labs(title="Histogram for number of clusters") +
    labs(x="Number of clusters", y="Density")
```

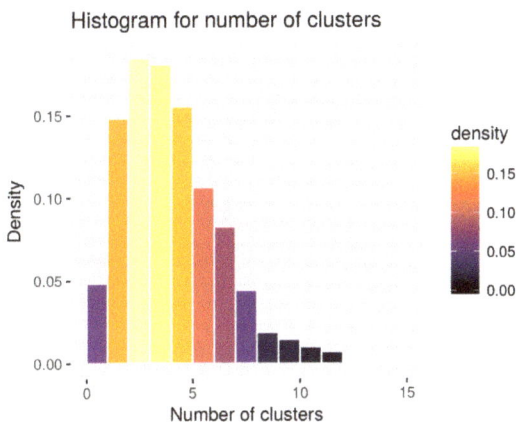

Histogram for number of clusters

The MAP estimator for the number of clusters is three, obtained as:

```
which.max(table(MCMC_number_cluster))
```

```
## 3
## 3
```

Note that the above histogram is also quite in agreement with an estimate of four clusters.

Stepping back, recall that a Bayes estimator is obtained from a formal decision theory rule: given a loss function, a Bayes estimator minimizes the posterior expected loss. For instance, with Euclidean parameter spaces,

- the L^2, squared loss provides the posterior mean,

- the L^1, absolute loss provides the posterior median,

- the $0-1$ loss provides the mode a posteriori (MAP).

We focus now on a loss function L on clusterings. The posterior expected loss of clustering c', denoted by $L(c')$, is obtained by averaging the loss with respect to posterior distribution, over the set of all partitions of the integers $1, \ldots, n$ denoted by $\mathcal{A}_n$

$$L(c') = \sum_{c \in \mathcal{A}_n} L(c, c') p(c|\mathbf{x}),$$

and the decision is taken by choosing the best partition

$$\hat{c} = \arg \min_{c' \in \mathcal{A}_n} \sum_{c \in \mathcal{A}_n} L(c, c') p(c|\mathbf{x}).$$

Several losses have been considered in the literature:

- 0-1 loss (Rajkowski, 2016),

- Binder loss (Dahl, 2006),

- Variation of information (VI) (Wade and Ghahramani, 2018).

The 0-1 loss gives rise to the MAP estimator:

$$L_{0-1}(c') = \sum_{c \in \mathcal{A}_n} L_{0-1}(c, c') p(c|\mathbf{x}) = \sum_{c \in \mathcal{A}_n, c \neq c'} p(c|\mathbf{x}) = 1 - p(c'|\mathbf{x})$$

which is to say that the expected loss of c' is all the posterior mass except that of c'. So that it is easily minimized at the value c' which has maximum posterior weight:

$$\hat{c} = \arg \min_{c' \in \mathcal{A}_n} L_{0-1}(c') = \arg \max_{c' \in \mathcal{A}_n} p(c'|\mathbf{x}) := MAP.$$

Negative results by Rajkowski (2016) show that the mode a posteriori (MAP) is inconsistent.

Instead of the 0-1 loss, one can resort to the Variation of information (VI) which was devised by Meilă (2007) for clustering comparison. It stems from information theory, and compares information (in terms of Shannon entropy H) in two clusterings with information shared between the two clusterings (I), see Meilă (2007) for details:

$$\mathrm{VI}(c, \hat{c}) = H(c) + H(\hat{c}) - 2I(c, \hat{c}).$$

Estimation under the Variation of information loss function was recently studied by Wade and Ghahramani (2018), with the interesting finding that this loss function tends to reduce the over-estimation of the number of cluster that is commonly obtained under Binder loss, for instance by Dahl (2006). See also Arbel et al. (2018a) for a further comparison between Binder and VI on varying sample sizes. We implement the Variation of information approach by using the mc-clust.ext R package developed by Wade and Ghahramani (2018). This requires to compute the posterior similarity matrix associated to the MCMC output, that is a matrix whose entries represent the posterior probability that two observations are clustered together (or rather, a Monte Carlo approximation of it).

```
posterior_similarity_matrix = comp.psm(MCMC_clustering)
```

It can be represented as follows, where darker colors depict a higher posterior probability of shared clustering.

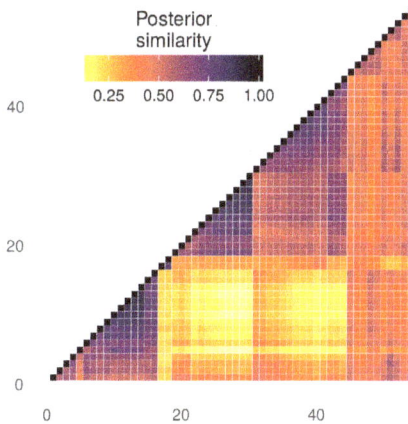

The next step consists in resolving the minimization problem

$$\hat{c} = \arg\min_{c' \in \mathcal{A}_n} \sum_{c \in \mathcal{A}_n} \mathrm{VI}(c, c') p(c|\mathbf{x}).$$

Of course, this optimization is only done approximately by the mcclust.ext R package, which scans the MCMC output and a neighbouring region of it in order to choose the best clustering.

```
clustering_Binder=minbinder.ext(posterior_similarity_matrix, method="greed
table(clustering_Binder$cl)

##
##  1  2  3  4  5
## 16  6 29  2  1

clustering_VI=minVI(posterior_similarity_matrix, method = "greedy")
table(clustering_VI$cl)

##
##  1
## 54
```

Thus, the Binder loss produces an estimated clustering with 5 groups, whereas under the Variation of information, all the observations are gathered into a single cluster of size 54. The interpretation of these results is that under the conditions of the assumed Dirichlet process mixture model, data are not deemed heterogeneous enough to justify multiple clusters under VI loss, whereas they do under Binder loss. The estimated clusterings and the ones obtained by Fraix-Burnet et al. (2009) are represented below in a 3D-like fashion by using the `rgl` package function `plot3d`.

```
subtitle = c("Binder loss, 5 groups", "VI loss, 1 group",
"Fraix-Burnet (2009), 4 groups", "Fraix-Burnet (2009), 3 groups")
clustering = cbind(clustering_Binder$cl,
                   clustering_VI$cl,
                   Grp3c,
                   Grp4c)
for(i in 1:4){
      cl = clustering[, i]
      plot3d(logTe, FeH, MV,
             type="s", size=2, col=plasma(max(cl))[cl],
             box = FALSE, sub = subtitle[i])
}
```

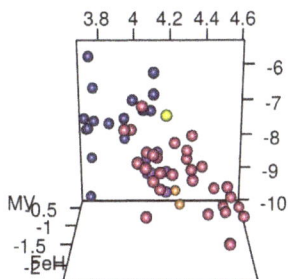

logTe
Binder loss, 5 groups

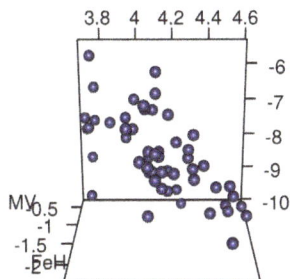

logTe
VI loss, 1 group

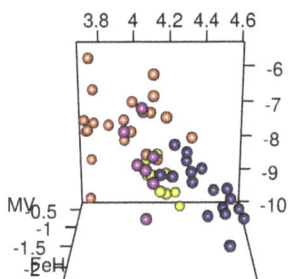

logTe
Fraix-Burnet (2009), 4 groups

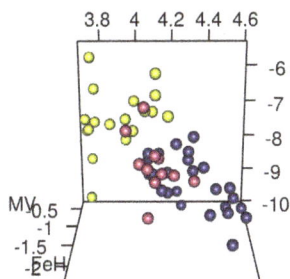

logTe
Fraix-Burnet (2009), 3 groups

Acknowledgements

I would like to thank Michał Lewandowski for helping with typing parts of this chapter and with merging the Globular data by Fraix-Burnet et al. (2009) with the Globular table that can be found on Wikipedia, see Footnote 2.

References

Aldous, D. J. (1985). Exchangeability and related topics. In *École d'Été de Probabilités de Saint-Flour XIII—1983*, pages 1–198. Springer.

Antoniak, C. E. (1974). Mixtures of Dirichlet processes with applications to Bayesian nonparametric problems. *The Annals of Statistics*, pages 1152–1174.

Arbel, J., Corradin, R., and Lewandowski, M. (2018a). Discussion of "Bayesian Cluster Analysis: Point Estimation and Credible Balls", by Wade and Ghahramani. *Bayesian Analysis*, 13(2).

Arbel, J., Corradin, R., and Nipoti, B. (2018b). Dirichlet process mixtures under affine transformations of the data. *Work in progress*.

Barrios, E., Lijoi, A., Nieto-Barajas, L. E., and Prünster, I. (2013). Modeling with normalized random measure mixture models. *Statistical Science*, 28(3):313–334.

Bernardo, J. M. and Smith, A. F. (2009). *Bayesian theory*, volume 405. Wiley.

Blackwell, D. and MacQueen, J. B. (1973). Ferguson distributions via Pólya urn schemes. *The Annals of Statistics*, pages 353–355.

Dahl, D. B. (2006). Model-based clustering for expression data via a Dirichlet process mixture model. *Bayesian inference for gene expression and proteomics*, pages 201–218.

Ferguson, T. (1973). A Bayesian analysis of some nonparametric problems. *The Annals of Statistics*, 1(2):209–230.

Fraix-Burnet, D., Davoust, E., and Charbonnel, C. (2009). The environment of formation as a second parameter for globular cluster classification. *Monthly Notices of the Royal Astronomical Society*, 398:1706–1714. To appear in MNRAS.

Jara, A., Hanson, T., Quintana, F., Müller, P., and Rosner, G. (2011). DPpackage: Bayesian non-and semi-parametric modelling in R. *Journal of statistical software*, 40(5):1.

Lo, A. (1984). On a class of Bayesian nonparametric estimates: I. Density estimates. *The Annals of Statistics*, 12(1):351–357.

Meilă, M. (2007). Comparing clusterings—an information based distance. *Journal of Multivariate Analysis*, 98(5):873–895.

Miller, J. W. and Harrison, M. T. (2013). A simple example of Dirichlet process mixture inconsistency for the number of components. In *Advances in neural information processing systems*, pages 199–206.

Miller, J. W. and Harrison, M. T. (2017). Mixture models with a prior on the number of components. *Journal of the American Statistical Association*, pages 1–17.

Müller, P. and Mitra, R. (2013). Bayesian nonparametric inference–why and how. *Bayesian Analysis*, 8(2).

Rajkowski, L. (2018). Analysis of the Maximal a Posteriori Partition in the Gaussian Dirichlet Process Mixture Model. *Bayesian Analysis*.

Sethuraman, J. (1994). A constructive definition of Dirichlet priors. *Statistica Sinica*, 4:639–650.

Wade, S. and Ghahramani, Z. (2018). Bayesian cluster analysis: Point estimation and credible balls. *Bayesian Analysis*, 13(2).

www.ingramcontent.com/pod-product-compliance
Lightning Source LLC
Chambersburg PA
CBHW042311210326
41598CB00041B/7348